LEGO® TRAIN PROJECTS

CHARLES PRITCHETT

**no starch
press**

San Francisco

CONTENTS

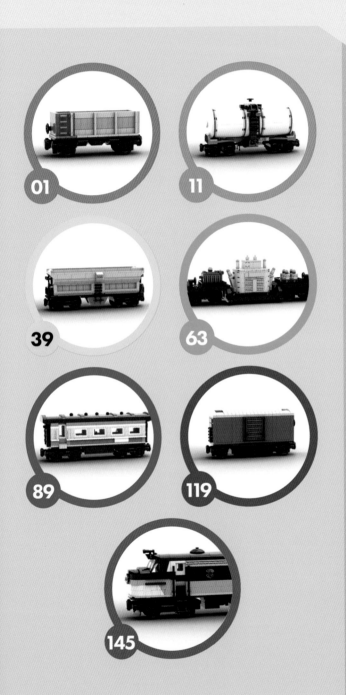

INTRODUCTION
SEEING THE LIGHT

It began in the mid-1980s when my mother purchased a large garbage bag full of used LEGO via a classified ad. In that bag was a treasure trove of amazing disassembled sets: early town, space, castle, and first-wave Technic sets, along with the original 1970s-era 12v train motor and a battery box.

Although the kind lady who sold the bag to my mother had also included a very large stack of instruction manuals, there was nothing train related. Armed with the latest LEGO Idea Book and the most recent catalogs, I set about trying to re-create their train designs from the pile of bricks in front of me. Watching the things I built actually move around the room on their blue tracks was amazing and brought building with LEGO to a new level of creativity for me.

Then, the Dark Ages.

As with most young people, my mind quickly turned to other interests. Girls, music, and video games took priority in my life, and LEGO fell by the wayside. My mother packed away the LEGO bricks. I graduated from high school, went to college, got married, had a daughter, and bought a house.

Then came the day that every former LEGO fanatic loves. My daughter was finally old enough to have her first LEGO set! As I sat with her and helped build Olivia's Invention Workshop (3933), my interest was rekindled. I realized what I had been missing for all those years. In short order, I was a collector and builder again—an Adult Fan of Lego (AFOL).

I dove back into the deep, armed with a little more time and a little more disposable income, and started buying the newest sets. There were now more than six colors! There were new elements I had never seen before! Bricks had multiple connection points and studs on the sides now! With so many new possibilities, there was no looking back.

Since then, I have gained another daughter to share the love of the hobby with, and in between helping build elves, dragons, and beasts, I've rediscovered my love of LEGO trains and their movement.

Keep on bricking!
Charles

HOW TO USE THIS BOOK

The instructions in this book should be very familiar to you if you've ever built an official LEGO retail set. You'll build the train cars in the official six-stud-wide format, and they should all be able to travel around your track without derailing. The models are built in steps and are mostly built from bottom to top.

This book presents each step with a small inset box showing the pieces and quantities that you'll need.

Whenever you'll need to flip or rotate before adding bricks, you'll see this icon.

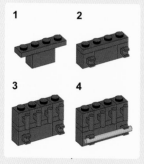

In some steps, you'll be required to complete submodels.

Complete the steps for building the submodel before attaching it to the main model.

FINDING PIECES

Each of the models in this book comes with a comprehensive bill of materials to help you find all the pieces you need before you start the building process. Start by checking your collection to see if you have the pieces needed. If, after you have exhausted your personal collection, you still don't have all of the pieces you require, you can buy the remaining ones either from LEGO itself or on the secondhand market.

Each LEGO retail store has a Pick-A-Brick wall that sells individual LEGO elements. If you aren't lucky enough to live near an official LEGO retail store, the LEGO Shop website (http://shop.lego.com) also has a Pick-A-Brick feature where you can purchase specific pieces in certain colors. LEGO does not, however, offer all the pieces and all the colors that have been available in the past, only those that are currently in production at its factories. For older pieces out of production, you will have to move to the secondhand market.

Bricklink (http://www.bricklink.com) is a website where you can choose from hundreds of sellers—and find just about anything that LEGO has ever released. Some of the pieces found in my instructions cannot be bought from LEGO itself any longer but can be found on Bricklink in plentiful quantities.

Before rushing out to buy all the pieces necessary, look over the instructions to see where certain pieces go. Some pieces are hidden deep inside the model, and their colors aren't written in stone. (They can be any color you have available.) Other times, you might want to change the color of something on the model. If the brick is visible on the outside of the model, using a color that is close will sometimes do the trick, such as using the older light gray instead of the newer light bluish gray color.

You also don't necessarily have to use the exact same pieces as I do in these instructions. Sometimes the number of studs and dimensions are far more important than the particular bricks used to achieve the effect. As long as a brick isn't structurally important to holding the model together, using other bricks of similar shape or size should work.

Digital versions of the bills of materials are available in Bricklink-compatible XML format for download at the following address:
http://www.brickmonster.toys/trains1

COAL GONDOLA

"Fill this car with some loose black 1×1 round studs for a coal load that looks great and adds texture!"

Pieces		144
Unique Bricks		36
Width	2.3 in	5.8 cm
Height	2.5 in	6.4 cm
Length	6.0 in	15.2 cm

1

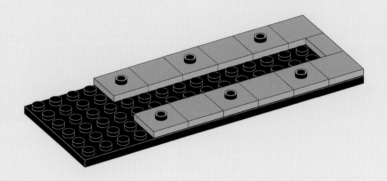

2

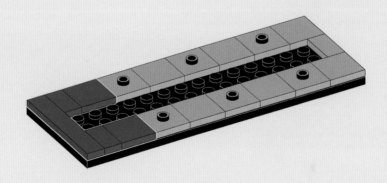

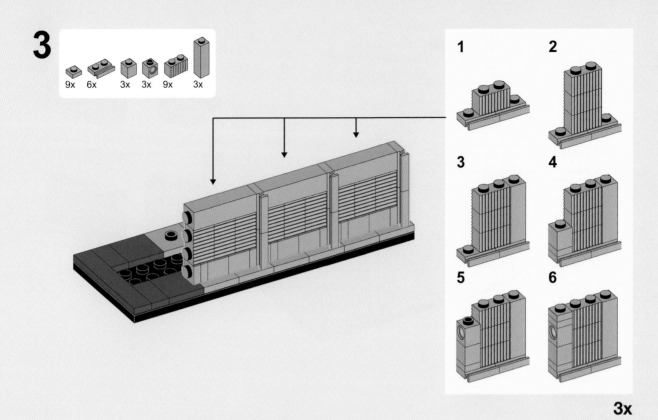

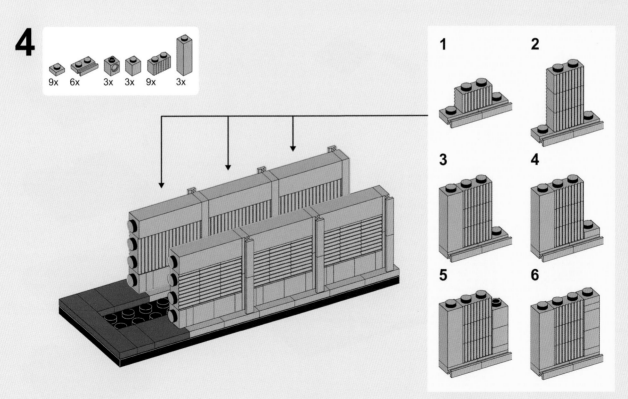

5

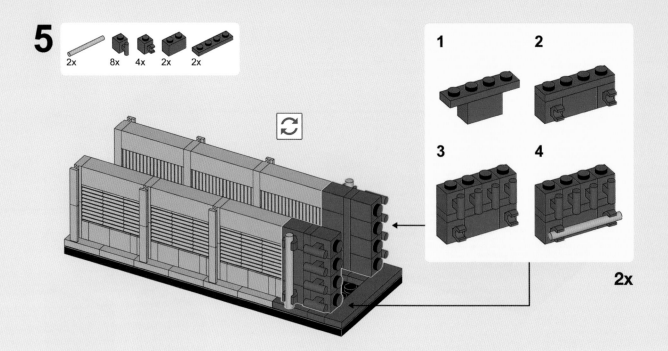

2x

6

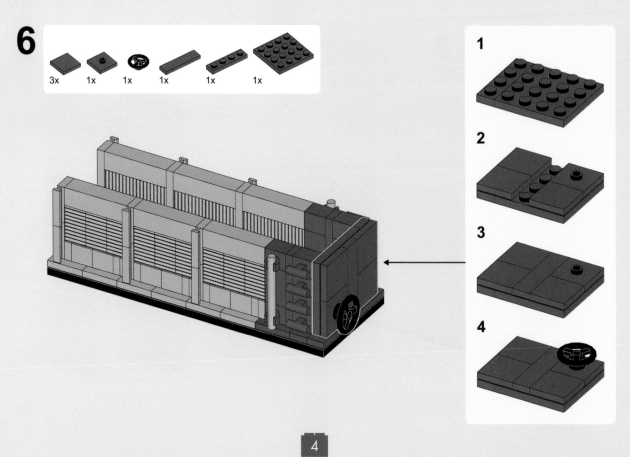

7

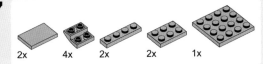

2x 4x 2x 2x 1x

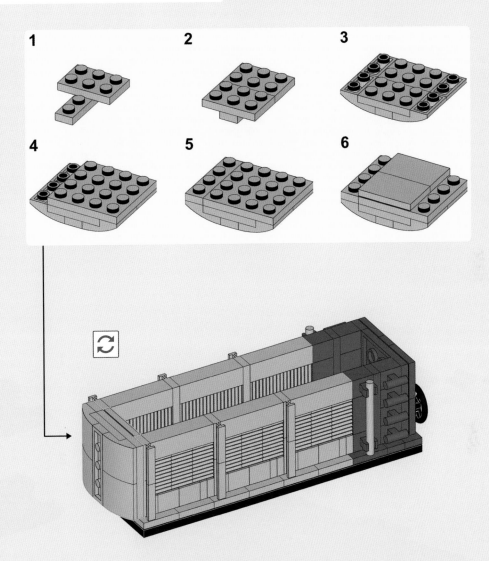

8

4x 4x

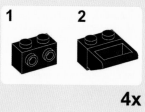

1 2

4x

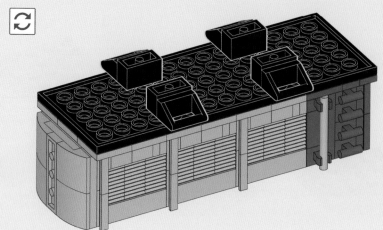

9

2x 1x 1x 2x

1

2

3

4

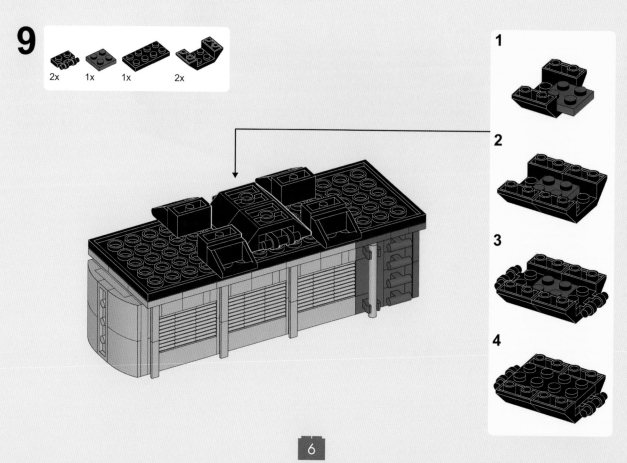

10

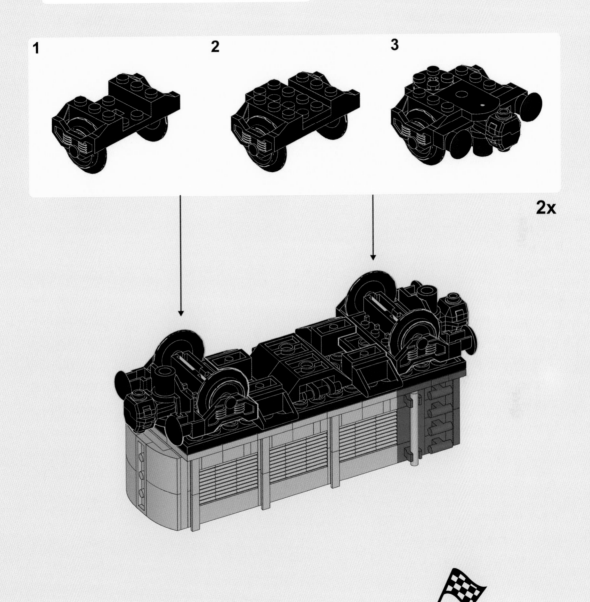

2x   2x   2x   2x

1   2   3

2x

COAL GONDOLA BILL OF MATERIALS

Item ID	Description	Color	Quantity
11211	Brick, Modified 1×2 with Studs on 1 Side	Black	4
3020	Plate 2×4	Black	3
3027	Plate 6×16	Black	1
2540	Plate, Modified 1×2 with Handle on Side - Free Ends	Black	2
4032	Plate, Round 2×2 with Axle Hole	Black	2
4871	Slope, Inverted 45° 4×2 Double	Black	2
2432	Tile, Modified 1×2 with Handle	Black	4
64424c01	Train Buffer Beam with Sealed Magnets - Type 1	Black	2
2878c02	Train Wheel RC Train, Holder with 2 Black Train Wheel RC Train and Chrome Silver Train Wheel RC Train, Metal Axle (2878 / 57878 / x1687)	Black	2
30663	Vehicle, Steering Wheel Small, 2 Studs Diameter	Black	1
30374	Bar 4L (Lightsaber Blade / Wand)	Light Bluish Gray	2
3005	Brick 1×1	Light Bluish Gray	6
14716	Brick 1×1×3	Light Bluish Gray	6
2877	Brick, Modified 1×2 with Grille (Flutes)	Light Bluish Gray	18
3024	Plate 1×1	Light Bluish Gray	18
3710	Plate 1×4	Light Bluish Gray	2
3021	Plate 2×3	Light Bluish Gray	2
3031	Plate 4×4	Light Bluish Gray	1
32028	Plate, Modified 1×2 with Door Rail	Light Bluish Gray	12
87580	Plate, Modified 2×2 with Groove and 1 Stud in Center (Jumper)	Light Bluish Gray	6
32803	Slope, Curved 2×2 Inverted	Light Bluish Gray	4
6541	Technic, Brick 1×1 with Hole	Light Bluish Gray	6
3069b	Tile 1×2 with Groove	Light Bluish Gray	1
3068b	Tile 2×2 with Groove	Light Bluish Gray	6
26603	Tile 2×3	Light Bluish Gray	2
3004	Brick 1×2	Red	2
60476	Brick, Modified 1×1 with Clip Horizontal	Red	4
2921	Brick, Modified 1×1 with Handle	Red	8
3710	Plate 1×4	Red	3
3022	Plate 2×2	Red	1
3031	Plate 4×4	Red	1
87580	Plate, Modified 2×2 with Groove and 1 Stud in Center (Jumper)	Red	1
3069b	Tile 1×2 with Groove	Red	2
2431	Tile 1×4	Red	1
6636	Tile 1×6	Red	1
3068b	Tile 2×2 with Groove	Red	5

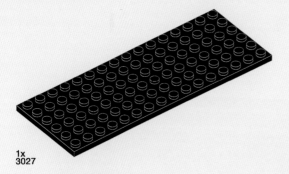

1x
3027

4x
2432

6x
6541

1x
3031

18x
3024

2x
64424c01

2x
3710

1x
6636

2x
30374

1x
3031

2x
2878c02

6x
14716

3x
3710

2x
3021

1x
2431

2x
4871

2x
26603

1x
3022

3x
3020

4x
32803

1x
87580

2x
4032

6x
87580

5x
3068b

1x
30663

6x
3068b

2x
3004

4x
11211

18x
2877

4x
60476

2x
2540

12x
32028

8x
2921

1x
3069b

2x
3069b

6x
3005

COAL GONDOLA ALTERNATE COLOR SCHEMES

MILK TANKER

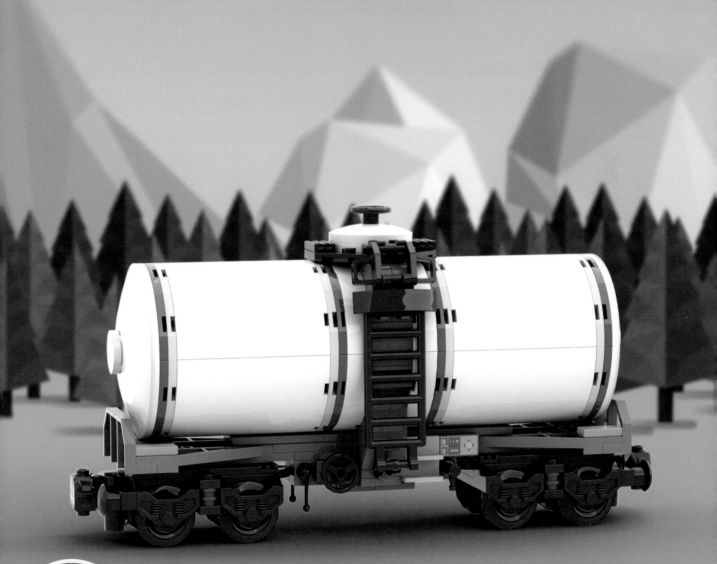

"By changing the colors of the stripes and tanks on this model, you can easily have any type of tanker your train needs!"

Pieces		188
Unique Bricks		63
Width	3.2 in	8.1 cm
Height	4.8 in	12.1 cm
Length	9.1 in	23.2 cm

1

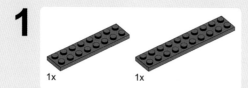

1x 1x

2

2x

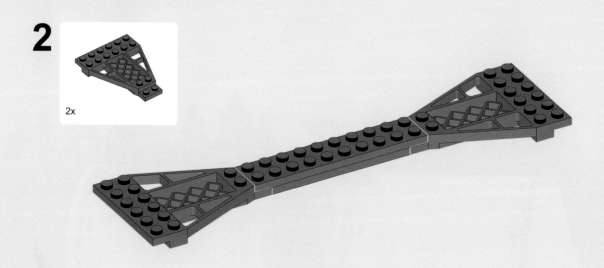

3

4x

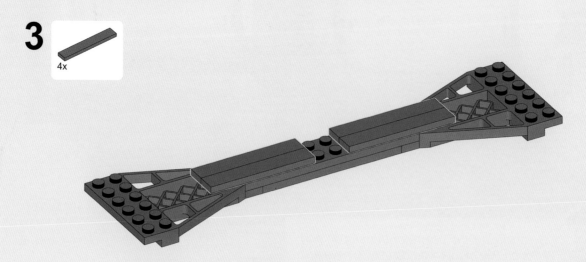

13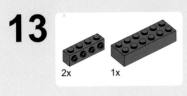

14

15

16

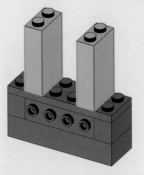

17

1x

18

4x

19

2x 1x

20

4x 2x

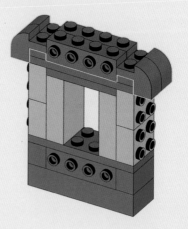

21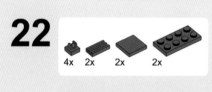
2x

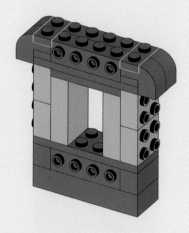

22

4x   2x   2x   2x

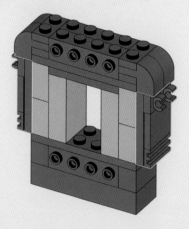

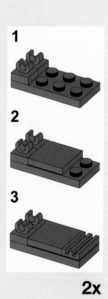

1

2

3

2x

23

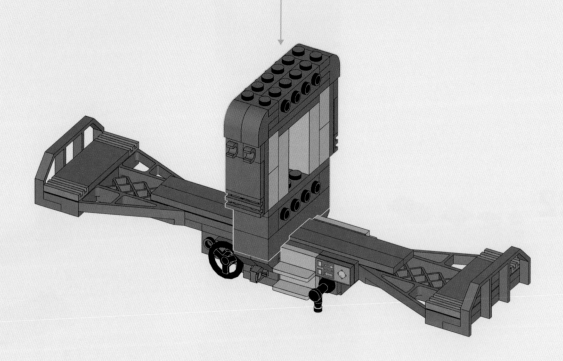

24
4x

25
2x

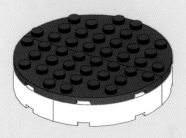

26
2x

27
1x   2x   1x

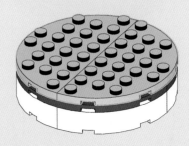

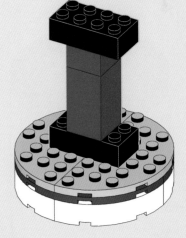

28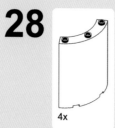

29

30

31

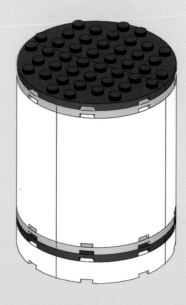

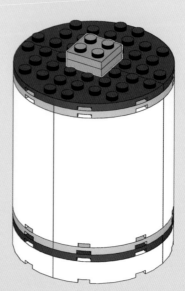

32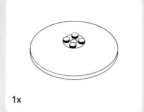

1x

33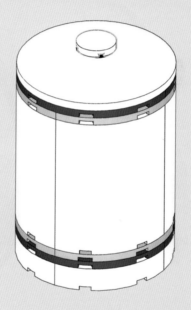

1x

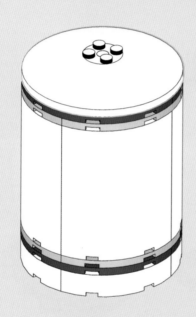

2x

34

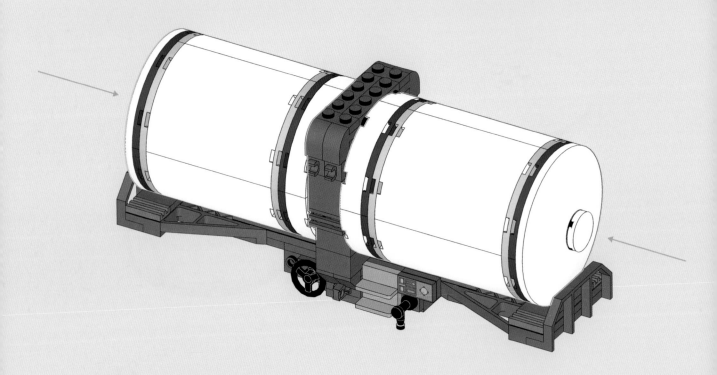

43

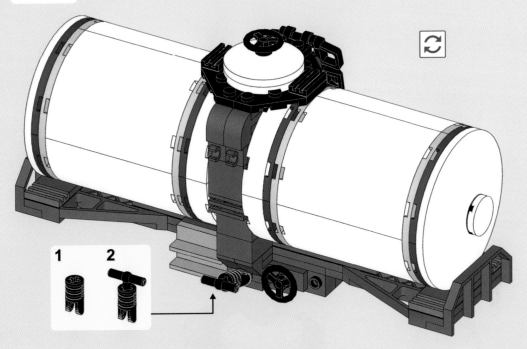

1 **2**

44

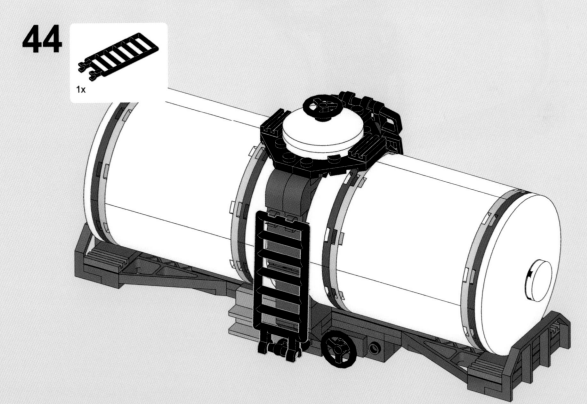

45

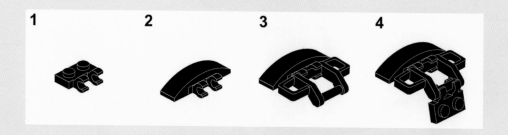

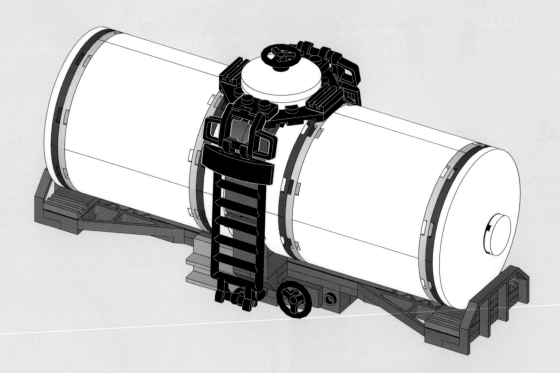

46

1x   1x

47

1x

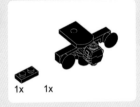

48

1x   1x

49

1x

50

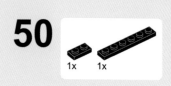

51

1 **2**

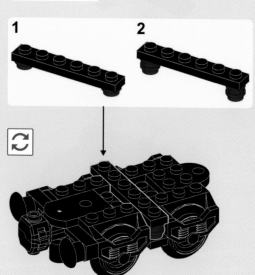

52

2x

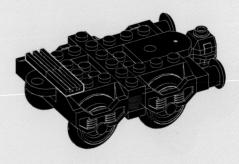

53

1x

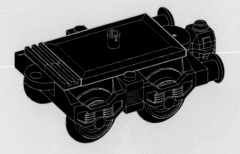

54

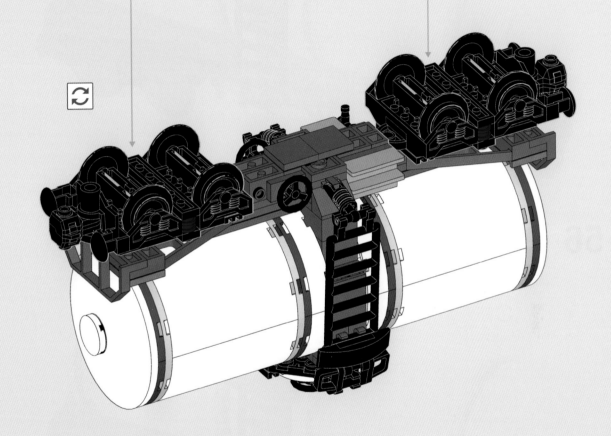

55

1x 1x

56

1x 1x

MILK TANKER BILL OF MATERIALS

Item ID	Description	Color	Quantity
35654	Bar 1×4×1 2/3 (Grille Guard / Push Bumper)	Black	2
6020	Bar 7×3 with Double Clips (Ladder)	Black	2
3001	Brick 2×4	Black	4
30553	Hinge Cylinder 1×2 Locking with 2 Fingers and Axle Hole on Ends	Black	2
4593	Lever Small	Black	2
4592	Lever Small Base	Black	2
3023	Plate 1×2	Black	4
3666	Plate 1×6	Black	4
3795	Plate 2×6	Black	2
60470b	Plate, Modified 1×2 with Clips Horizontal (thick open O clips)	Black	4
3176	Plate, Modified 3×2 with Hole	Black	2
4697b	Pneumatic T Piece Second Version (T Bar with Ball in Center)	Black	2
93273	Slope, Curved 4×1 Double	Black	2
4599b	Tap 1×1 without Hole in Nozzle End	Black	1
2412b	Tile, Modified 1×2 Grille with Bottom Groove / Lip	Black	6
4025	Train Bogie Plate (Tile, Modified 6×4 with 5mm Pin)	Black	2
64424c01	Train Buffer Beam with Sealed Magnets - Type 1	Black	2
2878c02	Train Wheel RC Train, Holder with 2 Black Train Wheel RC Train and Chrome Silver Train Wheel RC Train, Metal Axle (2878 / 57878 / x1687)	Black	4
30663	Vehicle, Steering Wheel Small, 2 Studs Diameter	Black	3
2419	Wedge, Plate 3×6 Cut Corners	Black	2
22888	Plate, Round Half 4×8	Blue	8
22888	Plate, Round Half 4×8	Bright Light Blue	8
3004	Brick 1×2	Dark Bluish Gray	2
2456	Brick 2×6	Dark Bluish Gray	1
87087	Brick, Modified 1×1 with Stud on 1 Side	Dark Bluish Gray	4
11211	Brick, Modified 1×2 with Studs on 1 Side	Dark Bluish Gray	4
6091	Brick, Modified 1×2×1 1/3 with Curved Top	Dark Bluish Gray	4
30414	Brick, Modified 1×4 with 4 Studs on 1 Side	Dark Bluish Gray	4
44567a	Hinge Plate 1×2 Locking with 1 Finger on Side with Bottom Groove	Dark Bluish Gray	2
3023	Plate 1×2	Dark Bluish Gray	8
3832	Plate 2×10	Dark Bluish Gray	1
3022	Plate 2×2	Dark Bluish Gray	2
2420	Plate 2×2 Corner	Dark Bluish Gray	1
3021	Plate 2×3	Dark Bluish Gray	1
3020	Plate 2×4	Dark Bluish Gray	4

Item ID	Description	Color	Quantity
3795	Plate 2×6	Dark Bluish Gray	2
3034	Plate 2×8	Dark Bluish Gray	1
3794	Plate, Modified 1×2 with 1 Stud without Groove (Jumper)	Dark Bluish Gray	2
6583	Plate, Modified 1×6 with Train Wagon End	Dark Bluish Gray	2
52501	Slope, Inverted 45 6×1 Double with 1×4 Cutout	Dark Bluish Gray	2
3069bpc1	Tile 1×2 with Groove with Vehicle Control Panel Pattern	Dark Bluish Gray	1
6636	Tile 1×6	Dark Bluish Gray	4
3068b	Tile 2×2 with Groove	Dark Bluish Gray	2
87079	Tile 2×4	Dark Bluish Gray	2
15712	Tile, Modified 1×1 with Clip - Rounded Edges	Dark Bluish Gray	4
2412b	Tile, Modified 1×2 Grille with Bottom Groove / Lip	Dark Bluish Gray	2
11203	Tile, Modified 2×2 Inverted	Dark Bluish Gray	2
30036	Wedge, Plate 8×6×2/3 with Grille	Dark Bluish Gray	2
85861	Plate, Round 1×1 with Open Stud	Dark Red	8
22886	Brick 1×2×3	Light Bluish Gray	2
22885	Brick, Modified 1×2×1 2/3 with Studs on 1 Side	Light Bluish Gray	4
4865b	Panel 1×2×1 with Rounded Corners	Light Bluish Gray	2
23950	Panel 1×3×1	Light Bluish Gray	2
3022	Plate 2×2	Light Bluish Gray	4
3003	Brick 2×2	Red	2
30145	Brick 2×2×3	Red	2
2577	Brick, Round Corner 4×4 Full Brick	White	8
30562	Cylinder Quarter 4×4×6	White	8
3960	Dish 4×4 Inverted (Radar) - Solid Stud	White	1
3961	Dish 8×8 Inverted (Radar) - Solid Studs	White	2
60474	Plate, Round 4×4 with Hole	White	1
14769	Tile, Round 2×2 with Bottom Stud Holder	White	2
2412b	Tile, Modified 1×2 Grille with Bottom Groove / Lip	Flat Silver	4

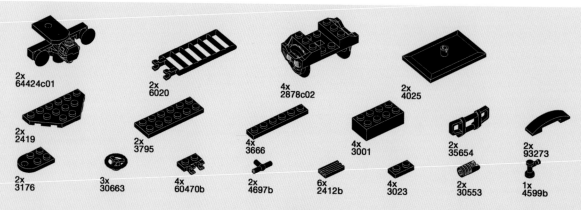

2x 64424c01
2x 6020
4x 2878c02
2x 4025
2x 2419
2x 3795
4x 3666
4x 3001
2x 35654
2x 93273
2x 3176
3x 30663
4x 60470b
2x 4697b
6x 2412b
4x 3023
2x 30553
1x 4599b

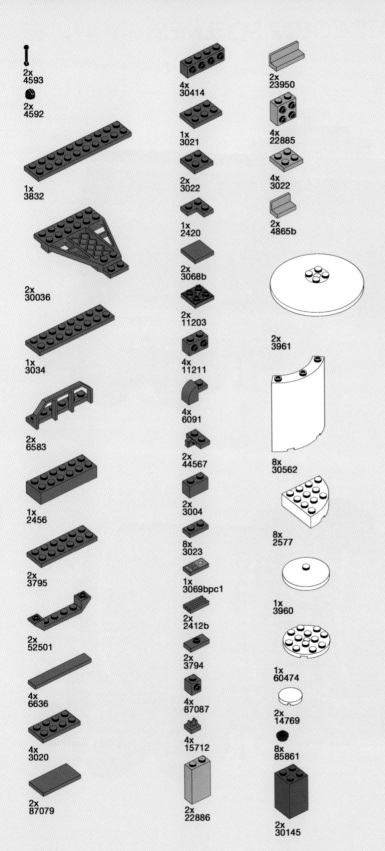

2x
4593

2x
4592

1x
3832

2x
30036

1x
3034

2x
6583

1x
2456

2x
3795

2x
52501

4x
6636

4x
3020

2x
87079

4x
30414

1x
3021

2x
3022

1x
2420

2x
3068b

2x
11203

4x
11211

4x
6091

2x
44567

2x
3004

8x
3023

1x
3069bpc1

2x
2412b

2x
3794

4x
87087

4x
15712

2x
22886

2x
23950

4x
22885

4x
3022

2x
4865b

2x
3961

8x
30562

8x
2577

1x
3960

1x
60474

2x
14769

8x
85861

2x
30145

2x
3003

8x
22888

8x
22888

4x
2412b

OPEN HOPPER

"Remember, you can use any colors that you have for these models. This one would look sharp with a red stripe too!"

Pieces		310
Unique Bricks		64
Width	3.3 in	8.3 cm
Height	3.6 in	9.2 cm
Length	9.8 in	24.8 cm

1

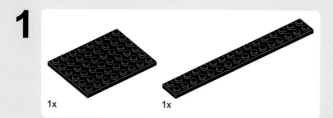

1x 1x

2

2x 2x

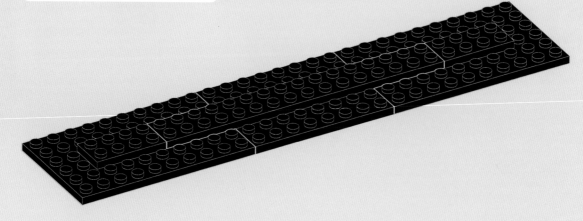

10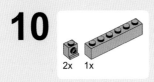

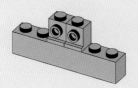

11

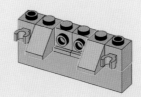

12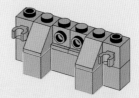

13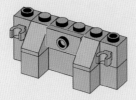

14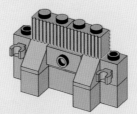

15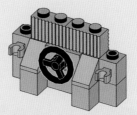

2x

16

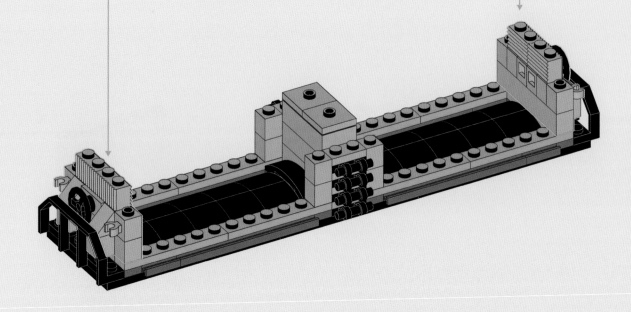

17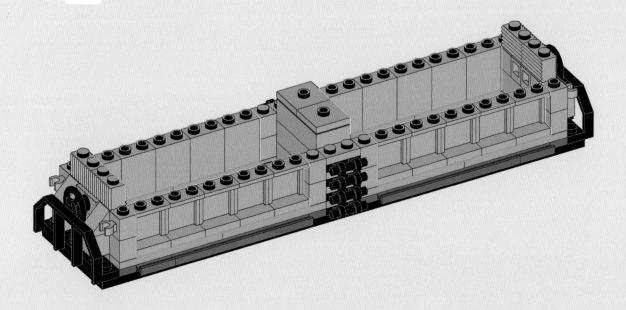

20x

18

24x　2x

19

8x　2x

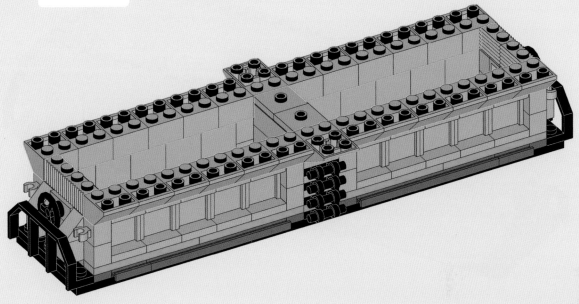

20

12x

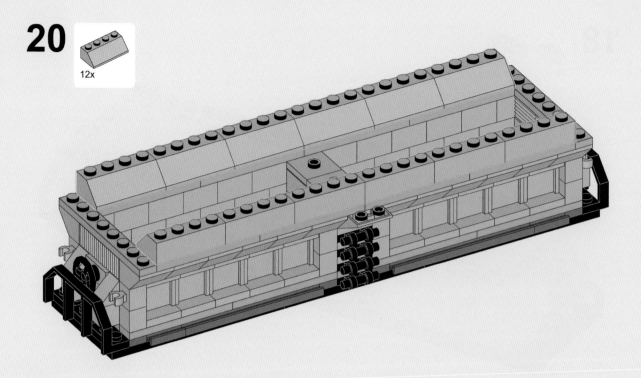

21

1x 2x 2x

1x

1 **2** **3**

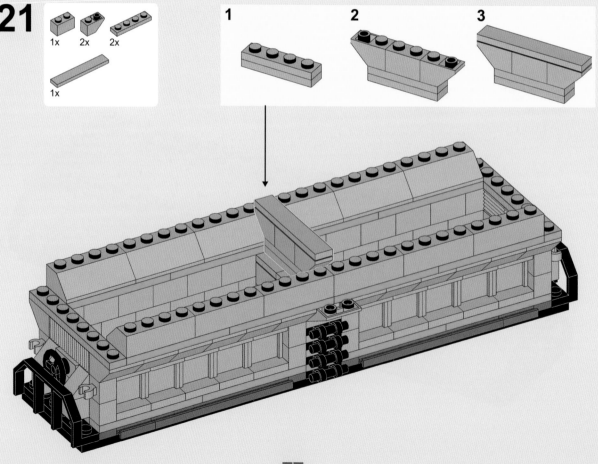

22

4x 2x

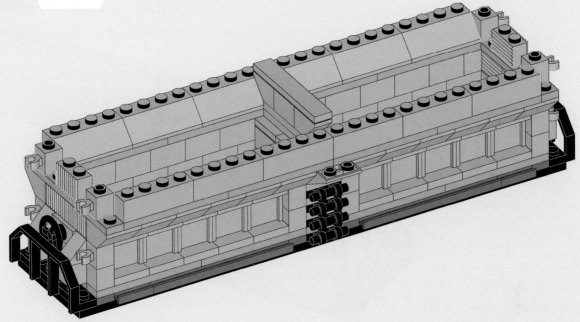

23

8x

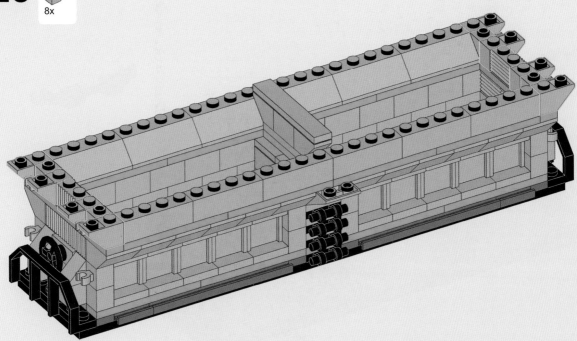

24

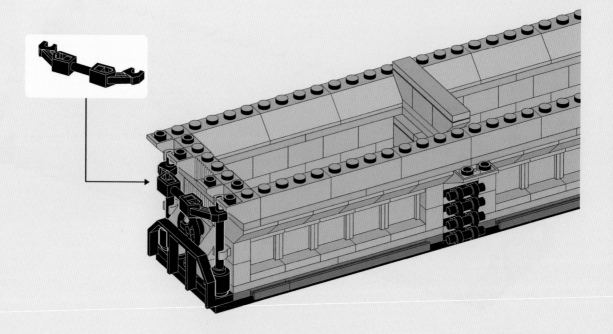

25

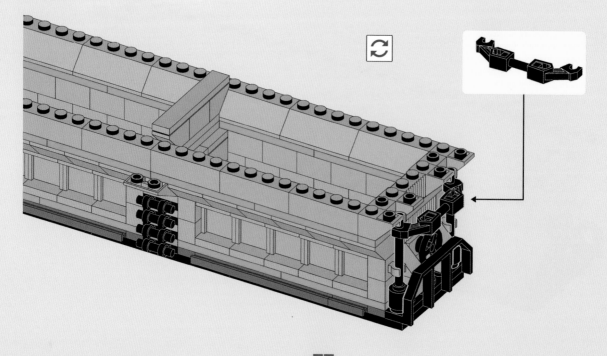

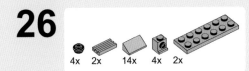

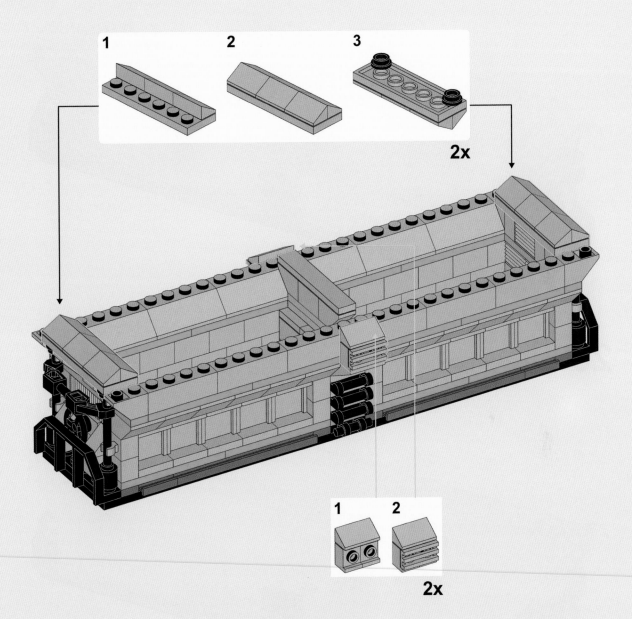

27

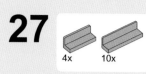

4x 10x

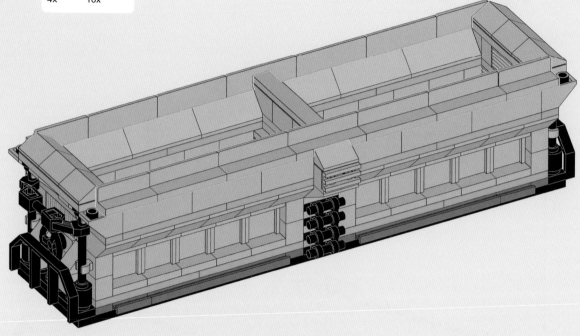

28

4x

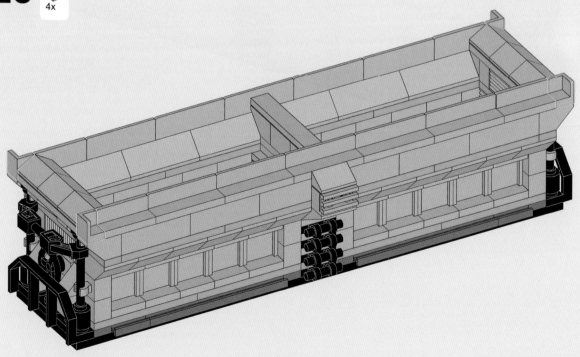

29

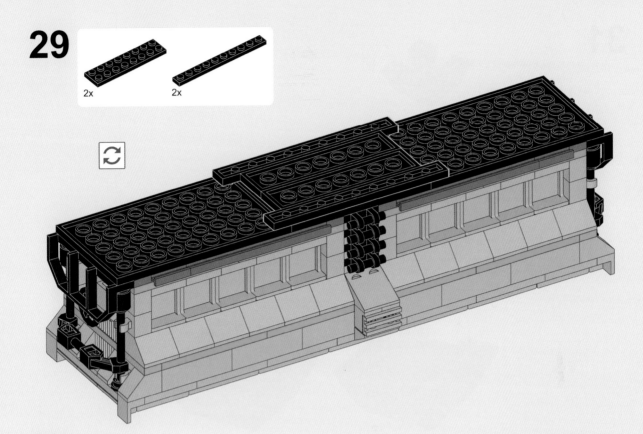

30

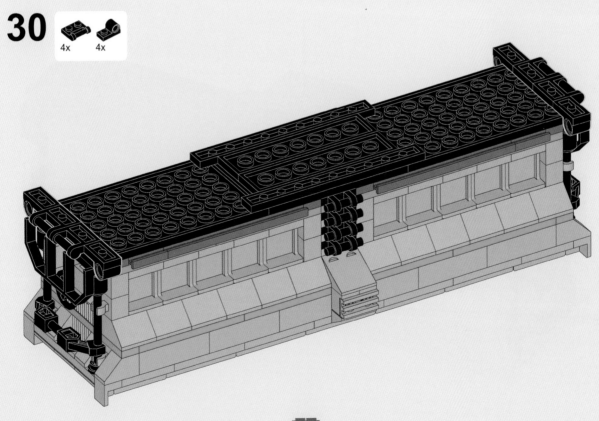

31

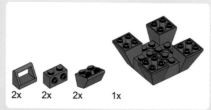

1 **2**

3 **4**

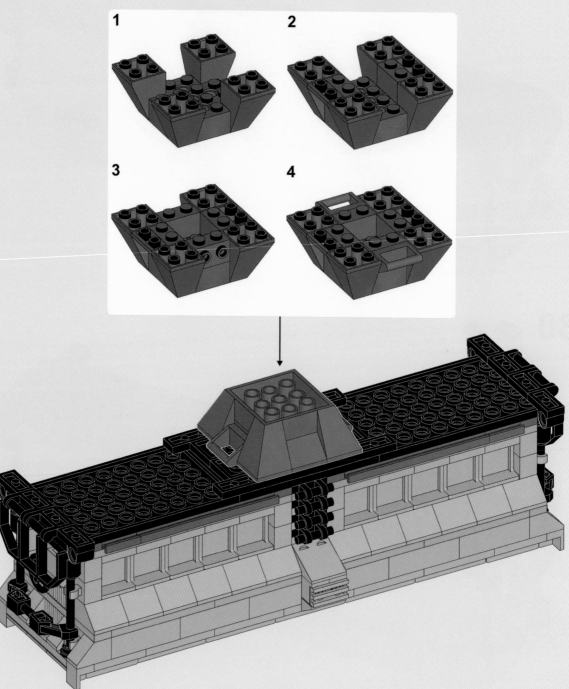

32

1x   1x

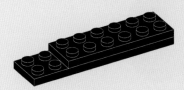

33

2x

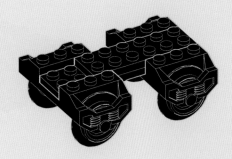

34

2x   2x   1x

1   **2**

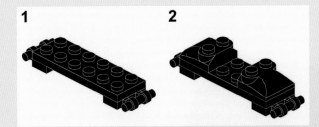

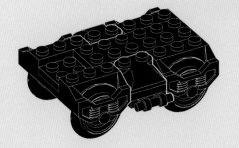

35

1x

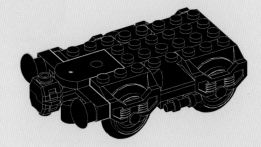

36

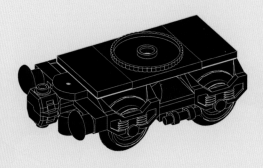

37

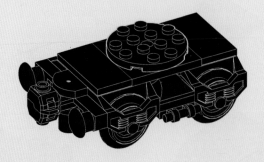

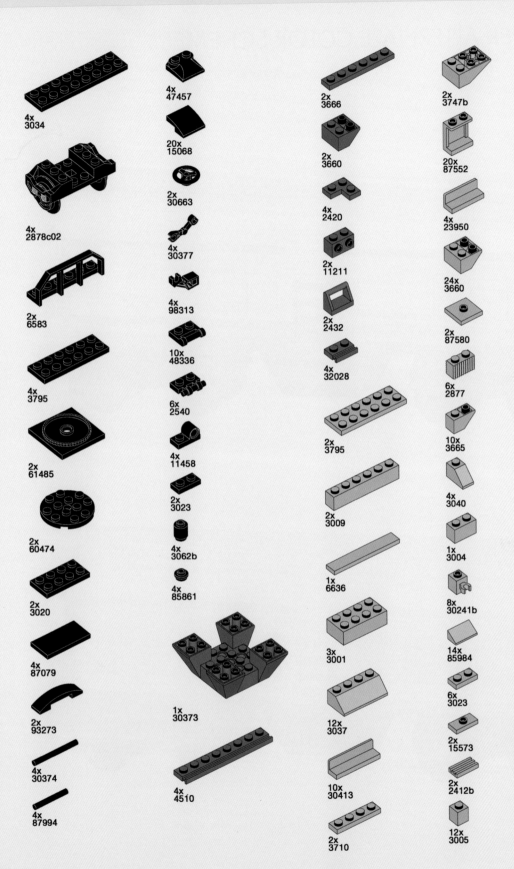

4x
3034

4x
2878c02

2x
6583

4x
3795

2x
61485

2x
60474

2x
3020

4x
87079

2x
93273

4x
30374

4x
87994

4x
47457

20x
15068

2x
30663

4x
30377

4x
98313

10x
48336

6x
2540

4x
11458

2x
3023

4x
3062b

4x
85861

1x
30373

4x
4510

2x
3666

2x
3660

4x
2420

2x
11211

2x
2432

4x
32028

2x
3795

2x
3009

1x
6636

3x
3001

12x
3037

10x
30413

2x
3710

2x
3747b

20x
87552

4x
23950

24x
3660

2x
87580

6x
2877

10x
3665

4x
3040

1x
3004

8x
30241b

14x
85984

6x
3023

2x
15573

2x
2412b

12x
3005

8x
4070

4x
6231

2x
3460

8x
3795

8x
3666

2x
3710

2x
3023

OPEN HOPPER ALTERNATE COLOR SCHEMES

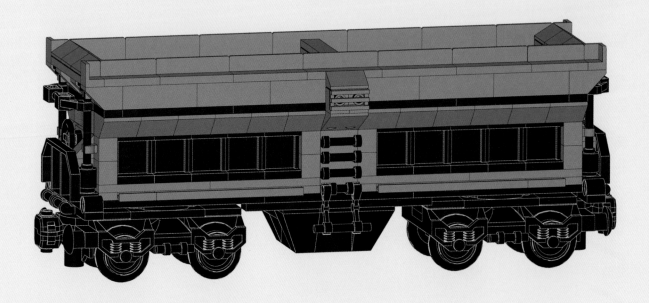

DEPRESSED FLATCAR
with ELECTRICAL LOAD

"The generator on this flatcar makes it pretty heavy, so make sure you have a strong locomotive to haul it!"

Pieces		382
Unique Bricks		92
Width	3.0 in	7.7 cm
Height	3.6 in	9.2 cm
Length	11.7 in	29.6 cm

1

2x 1x 1x

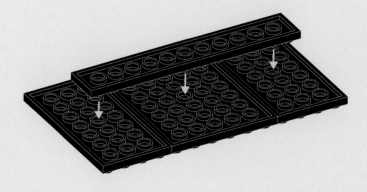

2

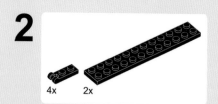

4x 2x

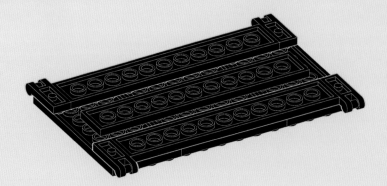

15

16

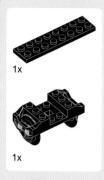

1x

1x

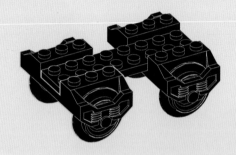

17

1x

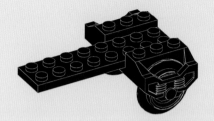

18

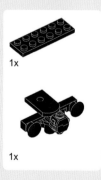

1x

1x

19

2x 2x 1x

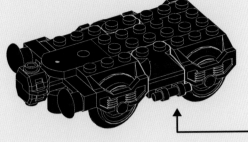

1

2

20

2x 1x

21

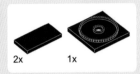

1x

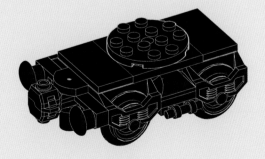

2x

22

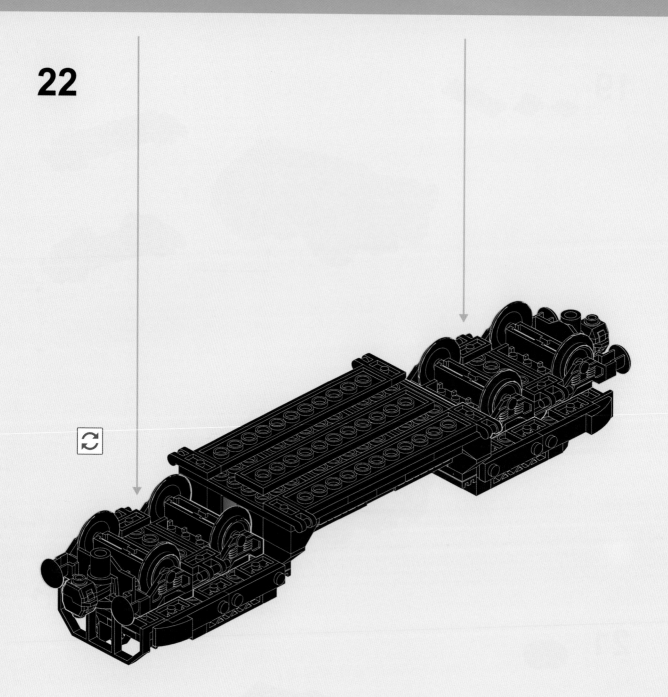

37

2x 2x 4x 1x 1x

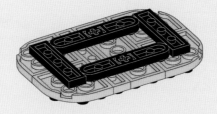

38

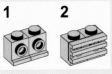

12x 24x

1 **2**

12x

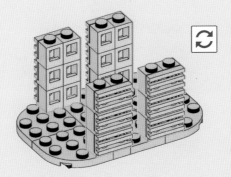

39

12x

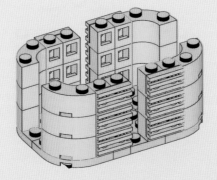

40

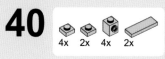

4x 2x 4x 2x

1 **2** **3** **4**

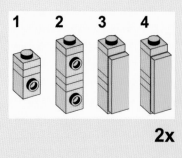

2x

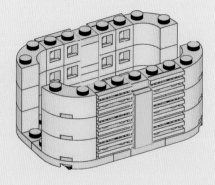

41

4x 4x 2x 4x

4x

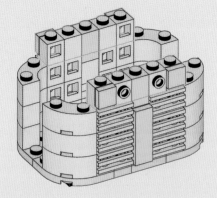

42

2x 4x

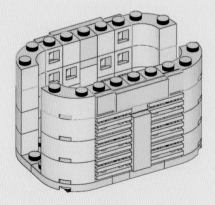

43  1x 1x 1x

44  1x

45  1x 2x

46 1x 1x

47  1x 1x

48  2x 2x

2x

49

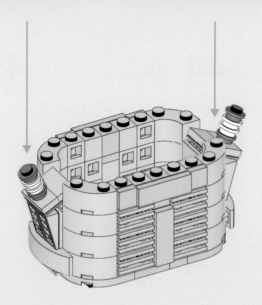

50

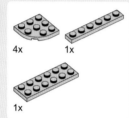

4x 1x

1x

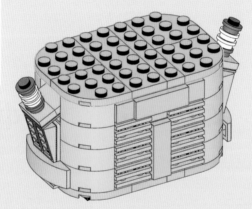

51

4x 4x 4x 2x

4x 2x

1x

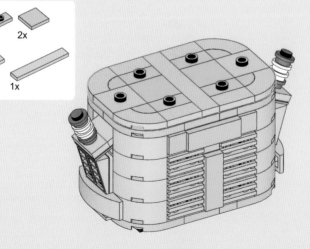

52

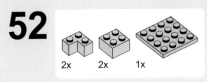

2x 2x 1x

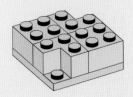

53

2x 2x

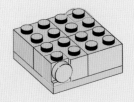

54

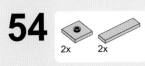

2x 2x

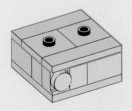

55

2x 2x

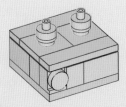

56

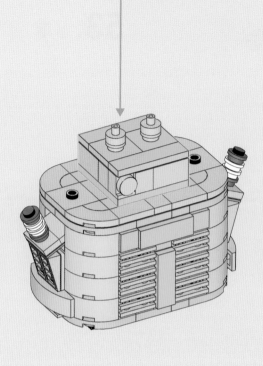

57 6x 4x

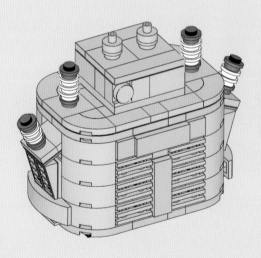

DEPRESSED FLATCAR BILL OF MATERIALS

Item ID	Description	Color	Quantity
99780	Bracket 1×2 - 1×2 Inverted	Black	12
47457	Brick, Modified 2×2×2/3 Two Studs, Curved Slope End	Black	4
3633	Fence 1×4×1	Black	12
6231	Panel 1×1×1 Corner	Black	8
4865b	Panel 1×2×1 with Rounded Corners	Black	2
3710	Plate 1×4	Black	2
2445	Plate 2×12	Black	3
3795	Plate 2×6	Black	4
3034	Plate 2×8	Black	2
3035	Plate 4×8	Black	2
3033	Plate 6×10	Black	2
3036	Plate 6×8	Black	1
60478	Plate, Modified 1×2 with Handle on End - Closed Ends	Black	4
2540	Plate, Modified 1×2 with Handle on Side - Free Ends	Black	4
92593	Plate, Modified 1×4 with 2 Studs without Groove	Black	8
6583	Plate, Modified 1×6 with Train Wagon End	Black	2
87580	Plate, Modified 2×2 with Groove and 1 Stud in Center (Jumper)	Black	2
4073	Plate, Round 1×1	Black	2
60474	Plate, Round 4×4 with Hole	Black	2
3040	Slope 45 2×1	Black	4
3039	Slope 45 2×2	Black	4
60481	Slope 65 2×1×2	Black	4
11477	Slope, Curved 2×1	Black	4
32530	Technic, Pin Connector Plate 1×2×1 2/3 with 2 Holes (Double on Top)	Black	2
32124	Technic, Plate 1×5 with Smooth Ends, 4 Studs and Center Axle Hole	Black	2
3069b	Tile 1×2 with Groove	Black	2
3068b	Tile 2×2 with Groove	Black	8
87079	Tile 2×4	Black	4
15712	Tile, Modified 1×1 with Clip - Rounded Edges	Black	4
2412b	Tile, Modified 1×2 Grille with Bottom Groove / Lip	Black	18
64424c01	Train Buffer Beam with Sealed Magnets - Type 1	Black	2
2878c02	Train Wheel RC Train, Holder with 2 Black Train Wheel RC Train and Chrome Silver Train Wheel RC Train, Metal Axle (2878 / 57878 / x1687)	Black	4
61485	Turntable 4×4 Square Base, Locking	Black	2
30663	Vehicle, Steering Wheel Small, 2 Studs Diameter	Black	2
3005	Brick 1×1	Bright Light Orange	2

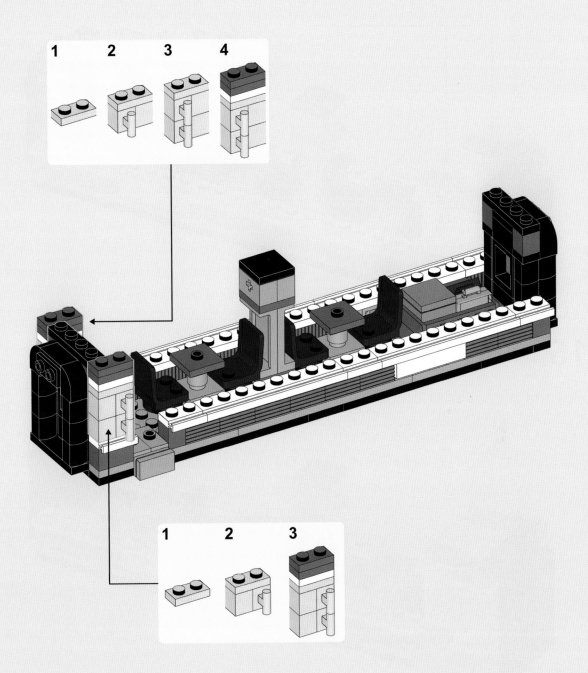

14

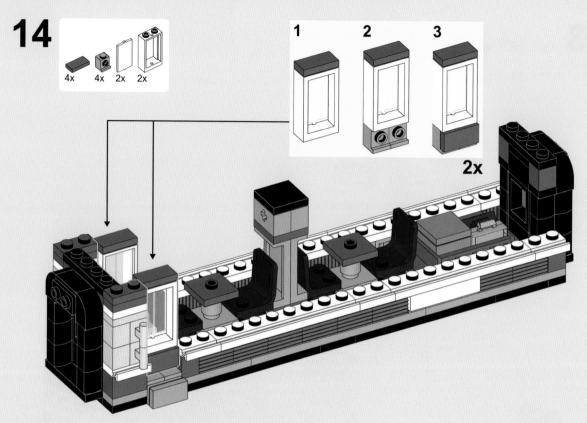

4x 4x 2x 2x

1   2   3

2x

15

8x

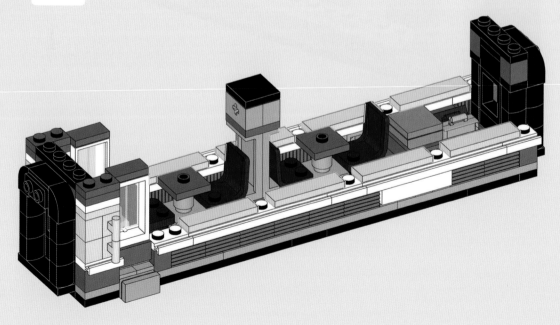

16

2x 2x 2x 4x

2x

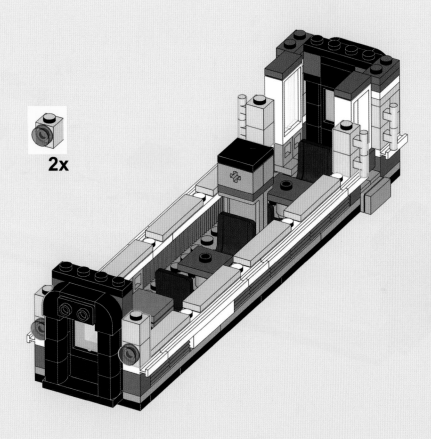

17

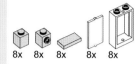

8x 8x 8x 8x 8x

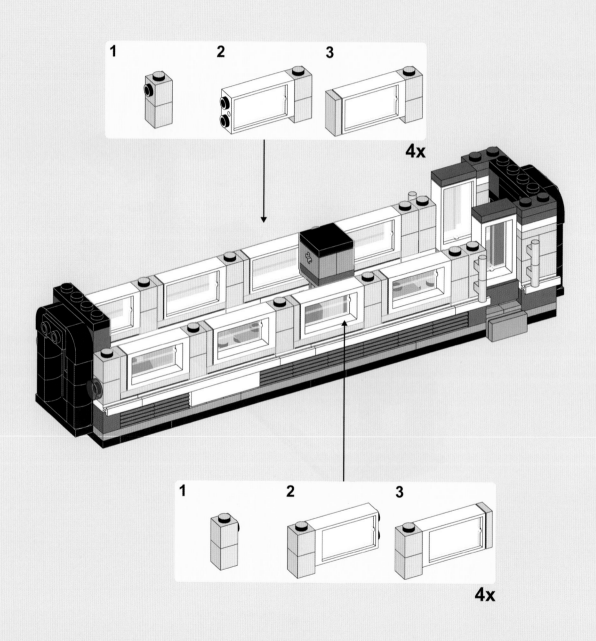

1 2 3 4x

1 2 3 4x

33

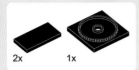

2x   1x

34

1x

2x

35

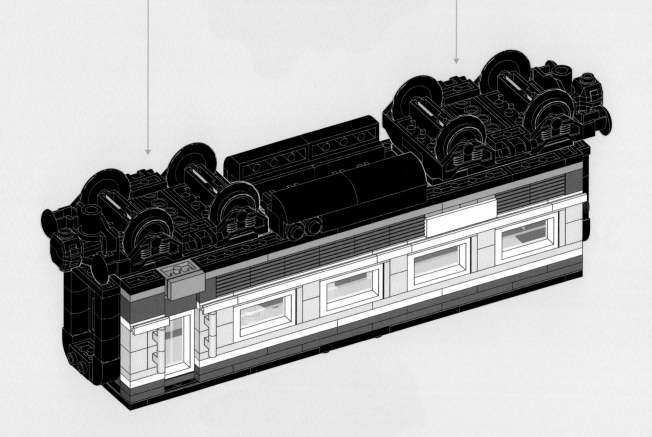

36

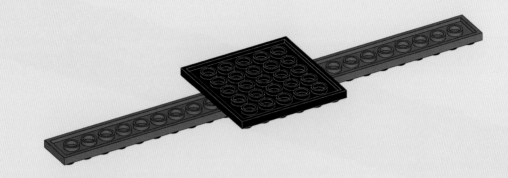

37

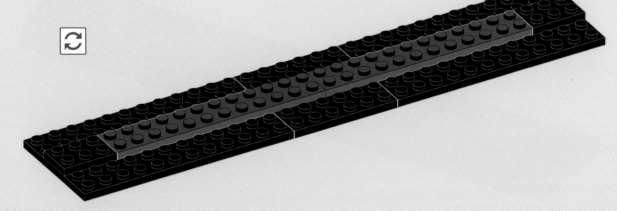

38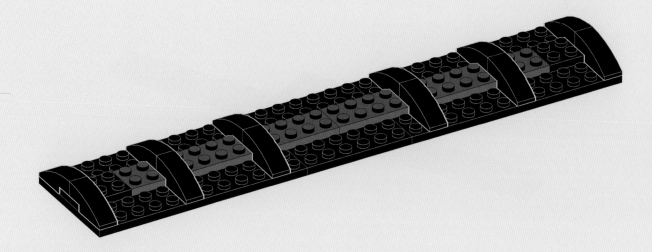

12x

39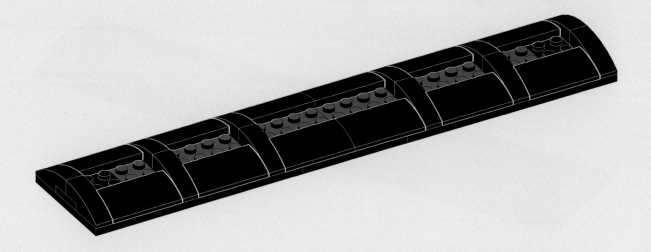

12x

40

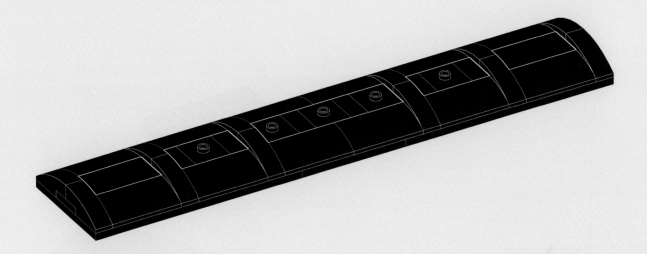

41

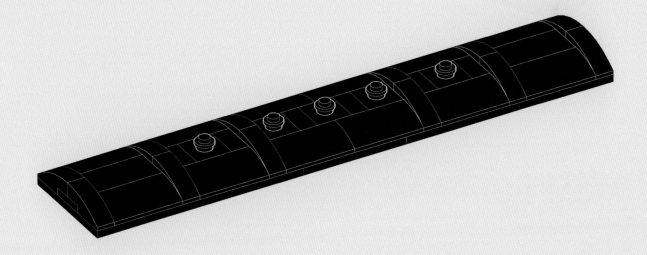

42

6x 2x

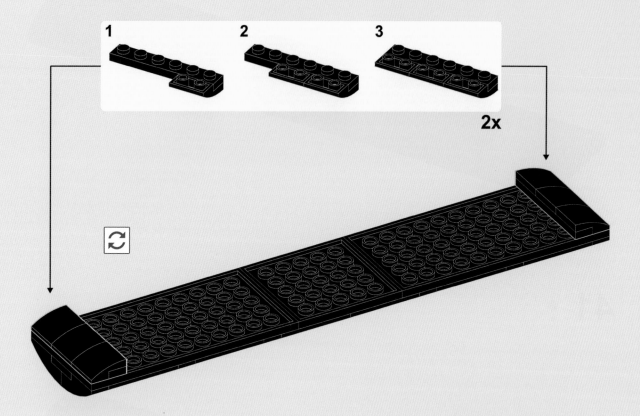

1   2   3

2x

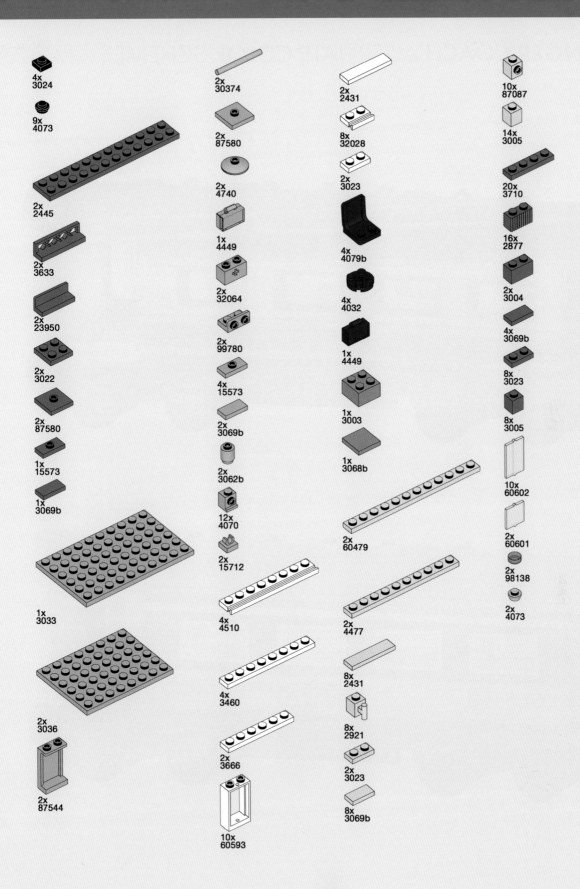

4x
3024

9x
4073

2x
2445

2x
3633

2x
23950

2x
3022

2x
87580

1x
15573

1x
3069b

1x
3033

2x
3036

2x
87544

2x
30374

2x
87580

2x
4740

1x
4449

2x
32064

2x
99780

4x
15573

2x
3069b

2x
3062b

12x
4070

2x
15712

4x
4510

4x
3460

2x
3666

10x
60593

2x
2431

8x
32028

2x
3023

4x
4079b

4x
4032

1x
4449

1x
3003

1x
3068b

2x
60479

2x
4477

8x
2431

8x
2921

2x
3023

8x
3069b

10x
87087

14x
3005

20x
3710

16x
2877

2x
3004

4x
3069b

8x
3023

8x
3005

10x
60602

2x
60601

2x
98138

2x
4073

POWERED
BOX CAR

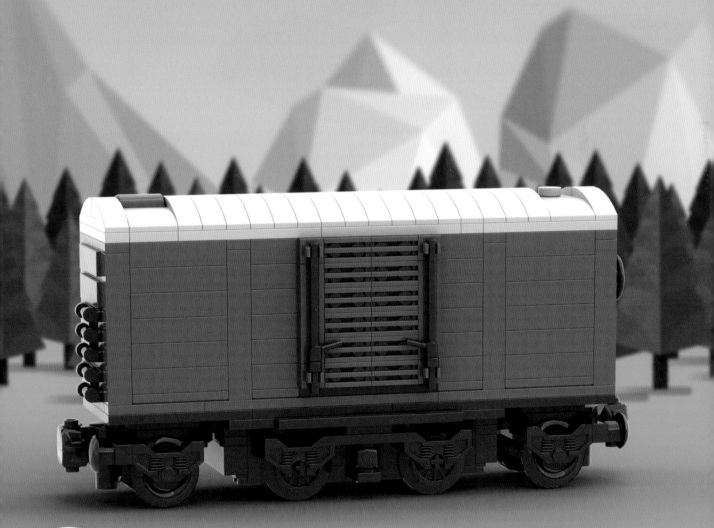

"With its integrated power functions, this car will help your train move along the tracks. If your train is especially long and heavy, you might need two!."

Pieces		144
Unique Bricks		36
Width	2.3in	5.8cm
Height	2.5in	6.4cm
Length	6.0in	15.2cm

1

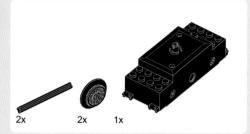

2x   2x   1x

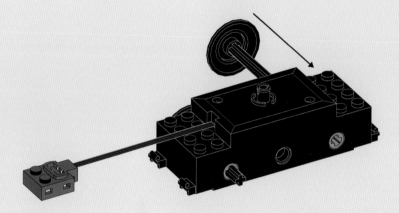

2

2x

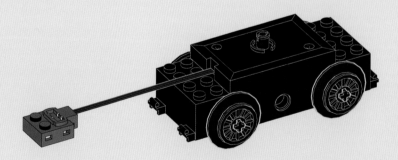

10

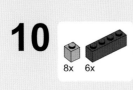

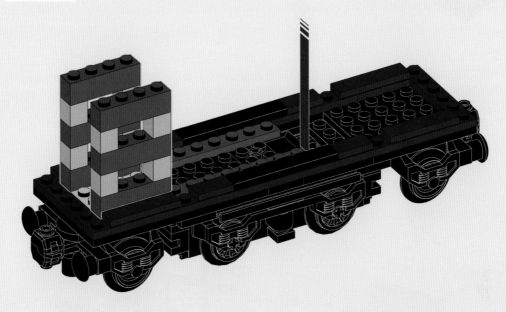

11

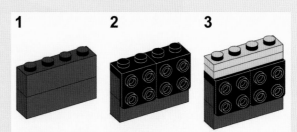

2x

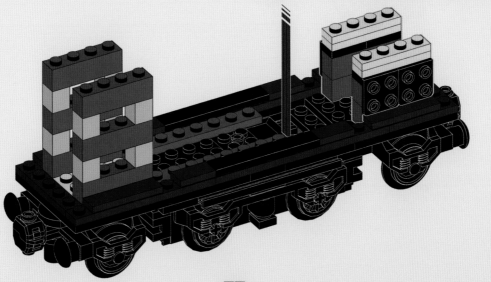

12

1x

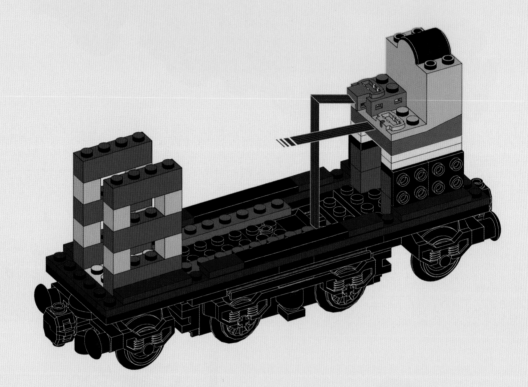

13

1x

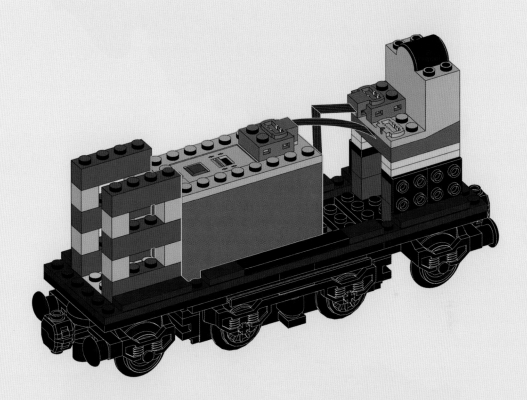

14 20x 2x

15 1x 4x 1x

16

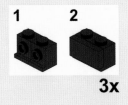

3x

17

18

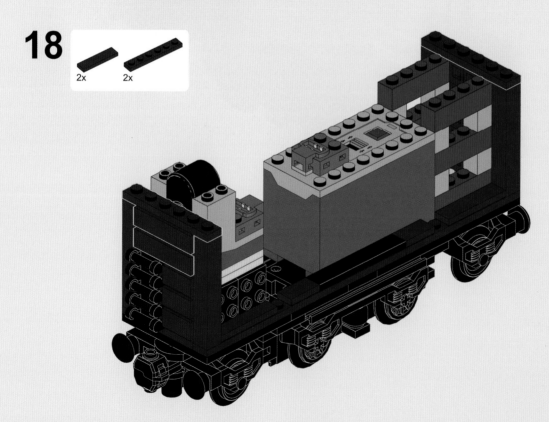

19

4x 4x

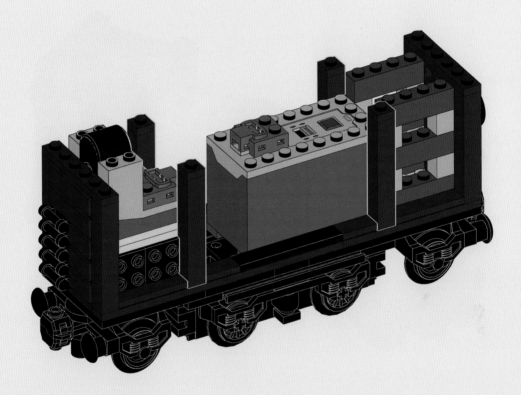

20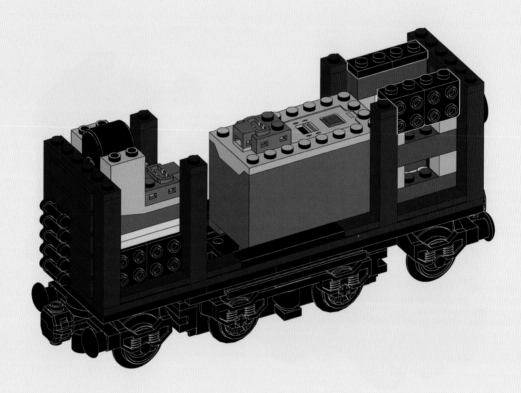

4x

21

8x 2x

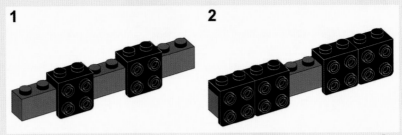

1　　　　　　　　**2**

2x

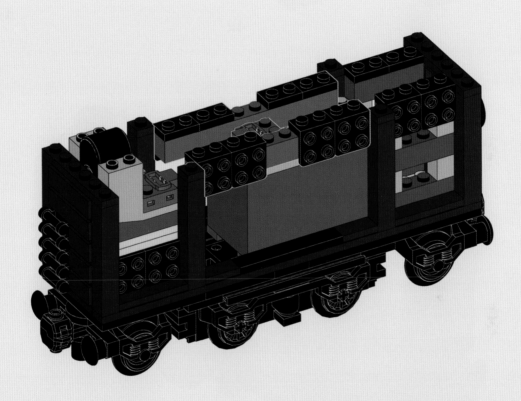

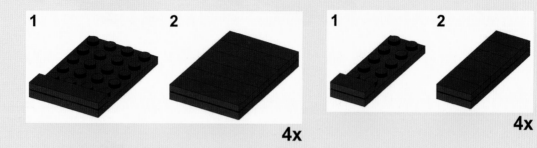

4x

4x

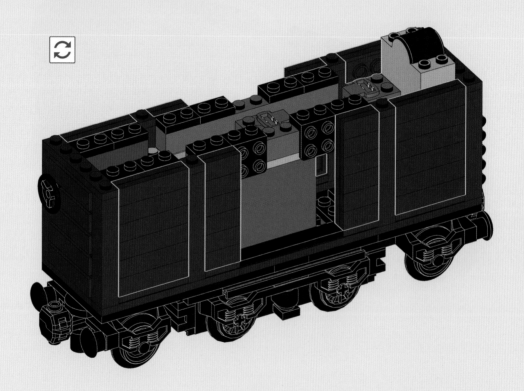

23
1x

24
2x

25
12x

 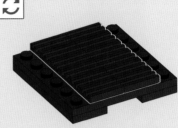

26
4x 2x

27
2x

28
2x

2x

29

2x 2x

1
2

2x

30

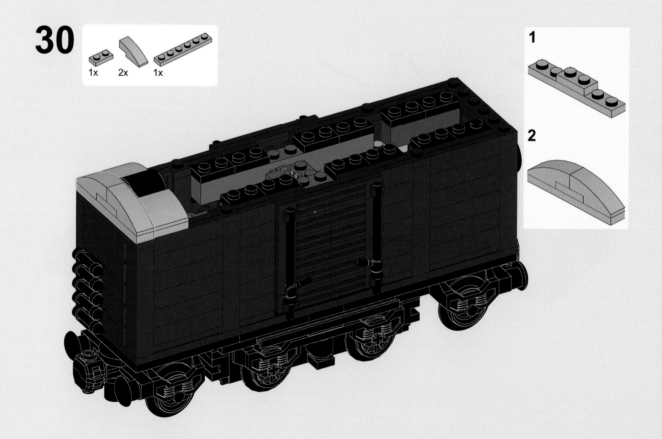

1x 2x 1x

1
2

Item ID	Description	Color	Quantity
3958	Plate 6×6	Light Bluish Gray	3
87580	Plate, Modified 2×2 with Groove and 1 Stud in Center (Jumper)	Light Bluish Gray	1
15068	Slope, Curved 2×2	Light Bluish Gray	4
50950	Slope, Curved 3×1	Light Bluish Gray	36
4274	Technic, Pin 1/2	Light Bluish Gray	2
98138	Tile, Round 1×1	Light Bluish Gray	1
3460	Plate 1×8	Red	2
2453a	Brick 1×1×5 - Blocked Open Stud or Hollow Stud	Reddish Brown	8
4070	Brick, Modified 1×1 with Headlight	Reddish Brown	34
3024	Plate 1×1	Reddish Brown	4
4477	Plate 1×10	Reddish Brown	2
3023	Plate 1×2	Reddish Brown	4
3710	Plate 1×4	Reddish Brown	4
3666	Plate 1×6	Reddish Brown	2
3460	Plate 1×8	Reddish Brown	2
3795	Plate 2×6	Reddish Brown	8
3032	Plate 4×6	Reddish Brown	4
3958	Plate 6×6	Reddish Brown	2
3069b	Tile 1×2 with Groove	Reddish Brown	31
63864	Tile 1×3	Reddish Brown	1
2431	Tile 1×4	Reddish Brown	38
15712	Tile, Modified 1×1 with Clip - Rounded Edges	Reddish Brown	8
2412b	Tile, Modified 1×2 Grille with Bottom Groove / Lip	Reddish Brown	24
4282	Plate 2×16	Tan	1
3710	Plate 1×4	Yellow	4
3020	Plate 2×4	Yellow	2

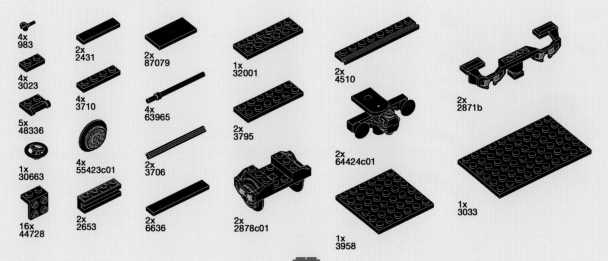

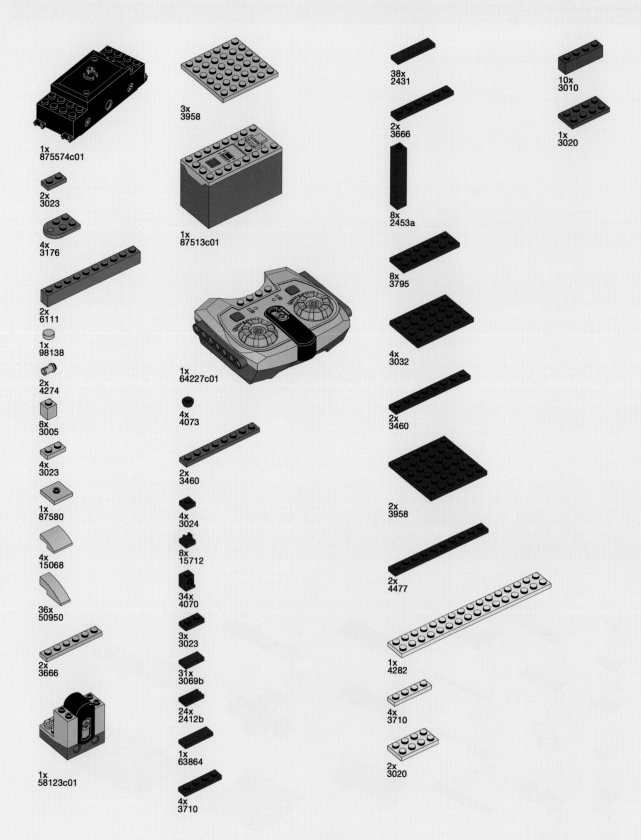

1x
875574c01

2x
3023

4x
3176

2x
6111

1x
98138

2x
4274

8x
3005

4x
3023

1x
87580

4x
15068

36x
50950

2x
3666

1x
58123c01

3x
3958

1x
87513c01

1x
64227c01

4x
4073

2x
3460

4x
3024

8x
15712

34x
4070

3x
3023

31x
3069b

24x
2412b

1x
63864

4x
3710

38x
2431

2x
3666

8x
2453a

8x
3795

4x
3032

2x
3460

2x
3958

2x
4477

1x
4282

4x
3710

2x
3020

10x
3010

1x
3020

POWERED BOXCAR ALTERNATE COLOR SCHEMES

EMD FL9 LOCOMOTIVE

"Since this locomotive doesn't have any power functions built in, you'll have to pair it with the Powered Box Car to get your train moving along the tracks."

Pieces		698
Unique Bricks		161
Width	2.8in	7.1cm
Height	4.3in	10.9cm
Length	14.3in	36.2cm

1

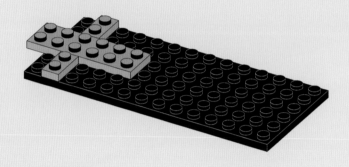

2

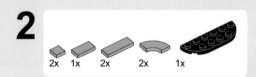

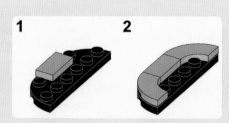

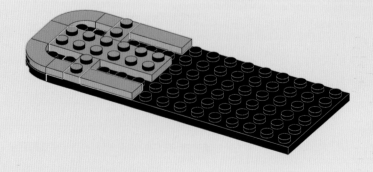

3

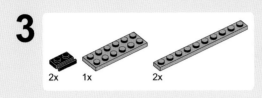

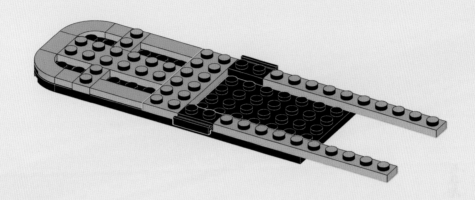

4

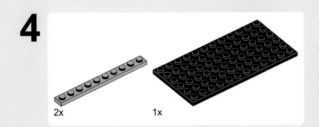

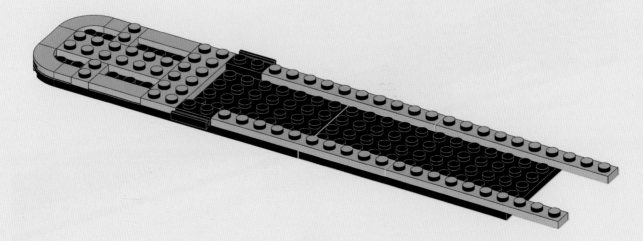

5

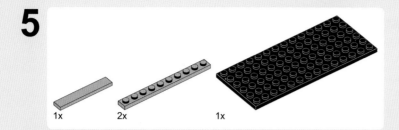

1x   2x   1x

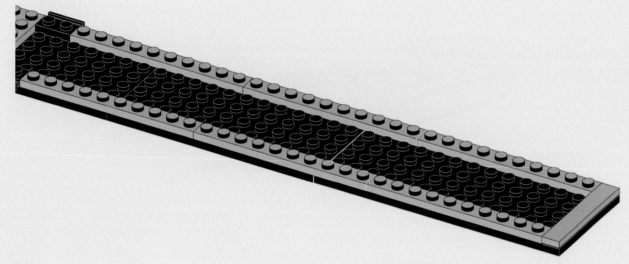

6

2x 2x

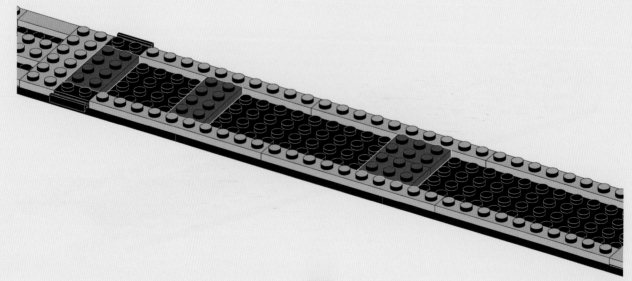

15

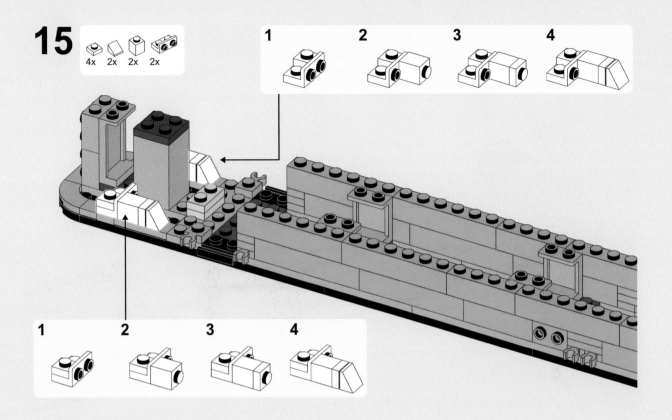

16

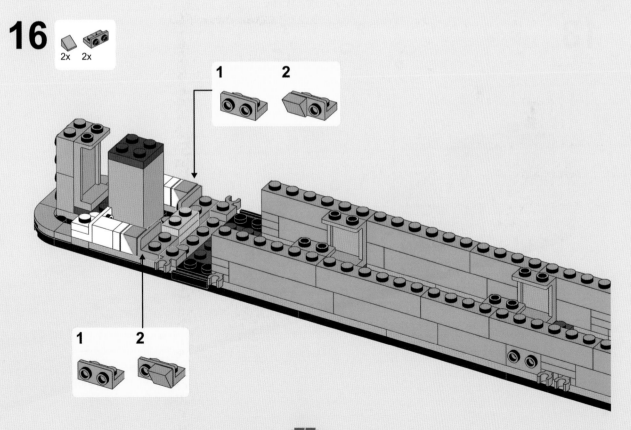

17

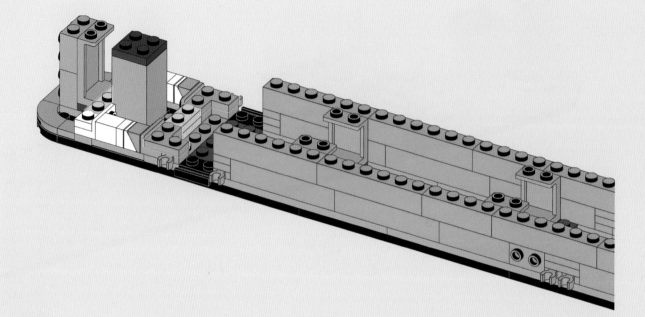

18

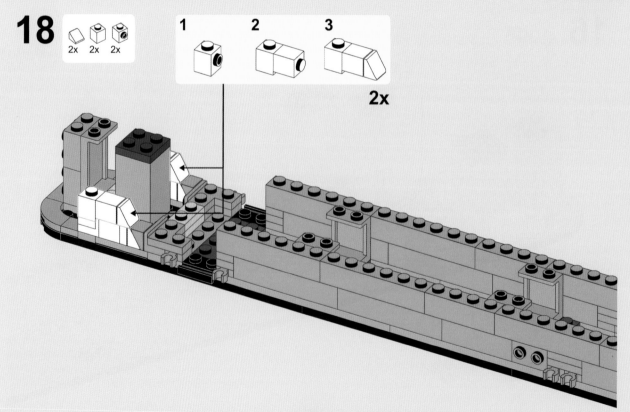

19

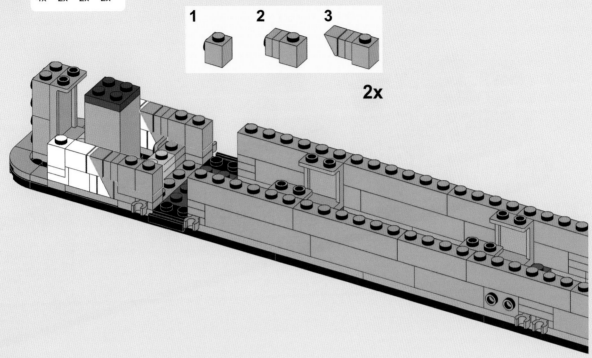

20

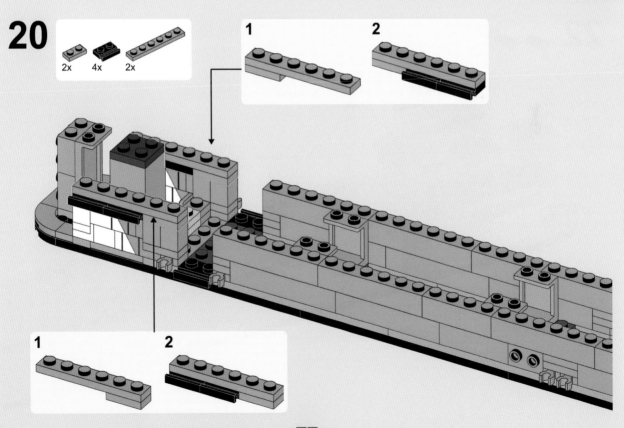

21

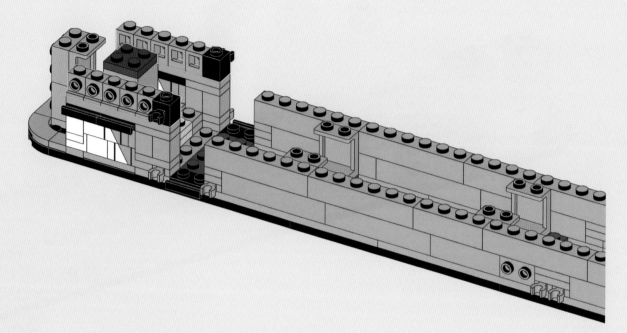

22

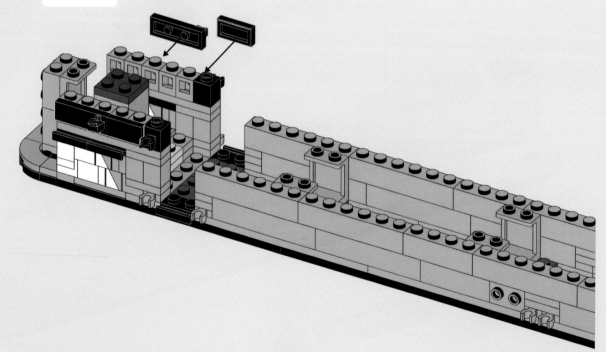

23

2x 1x 4x

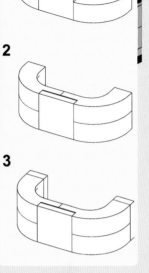

1

2

3

24

2x 1x

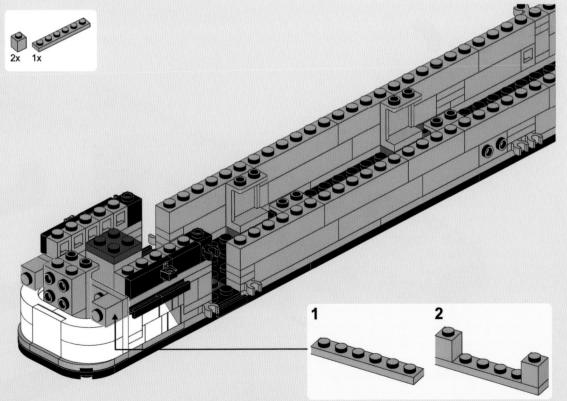

1

2

25

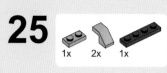

1x 2x 1x

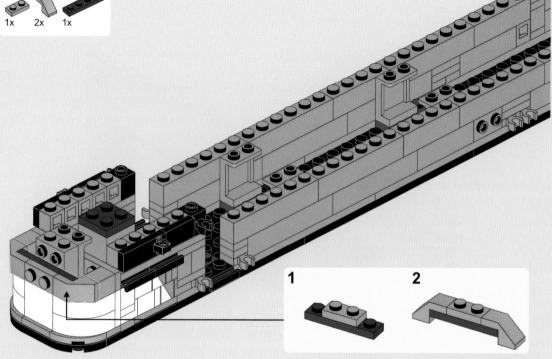

1

2

26

2x 1x 2x

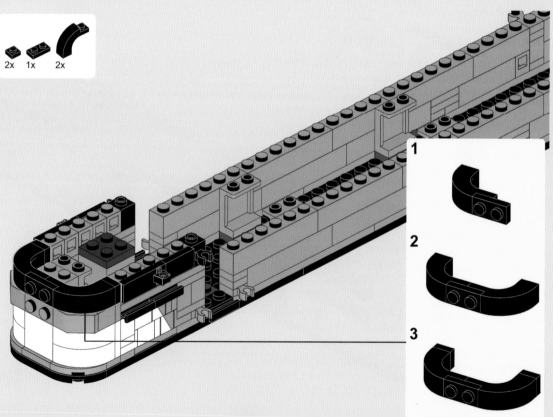

1

2

3

27

1x 1x

1 **2**

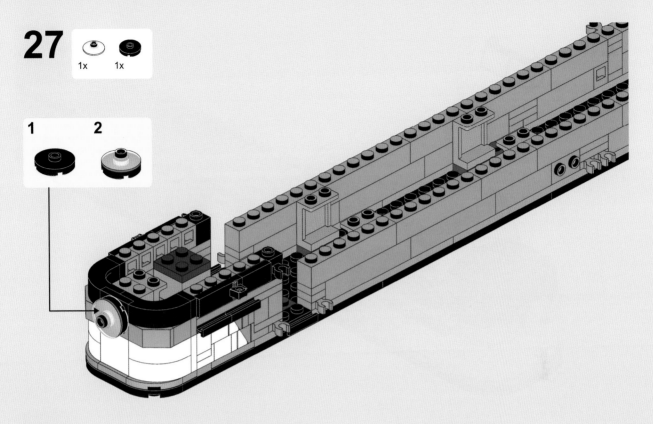

28

1x 1x 2x 1x 2x

1

2

3

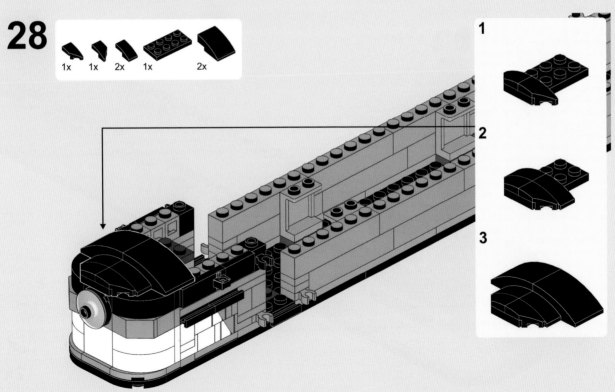

29

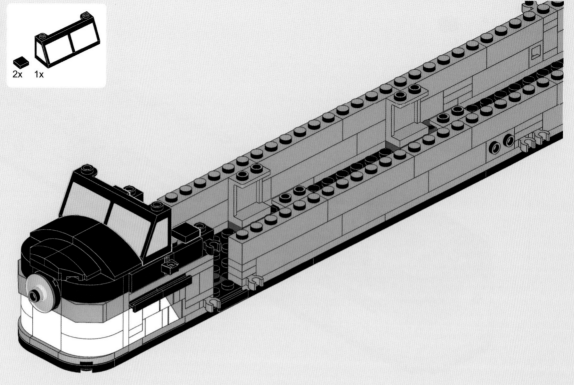

2x 1x

30

4x 2x 2x 2x 2x

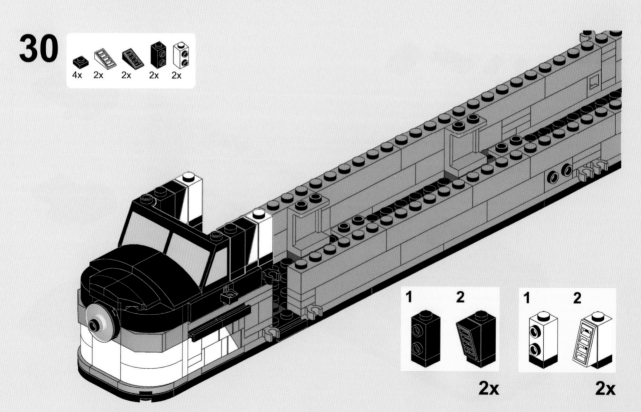

1   2

2x

1   2

2x

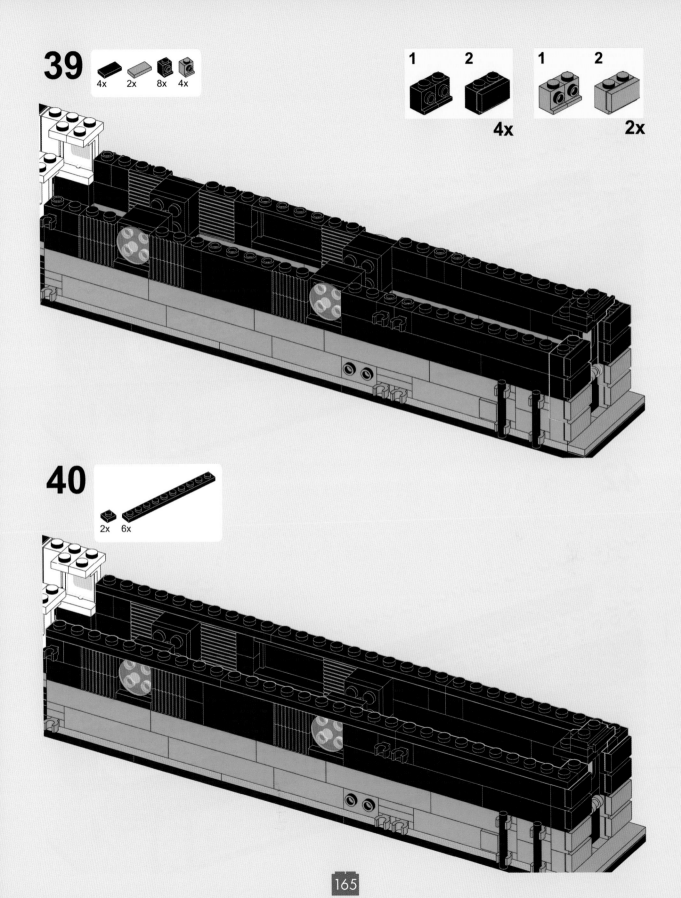

41

58x

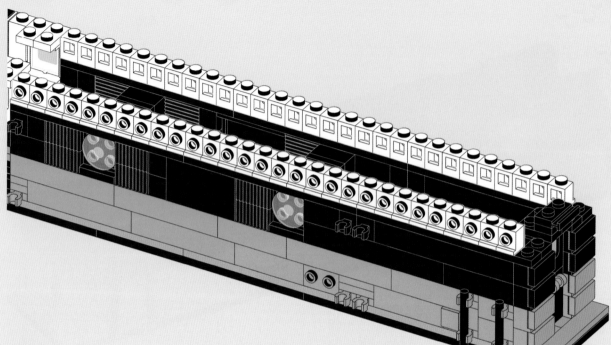

42

1x 2x 2x

| 1 | 2 |

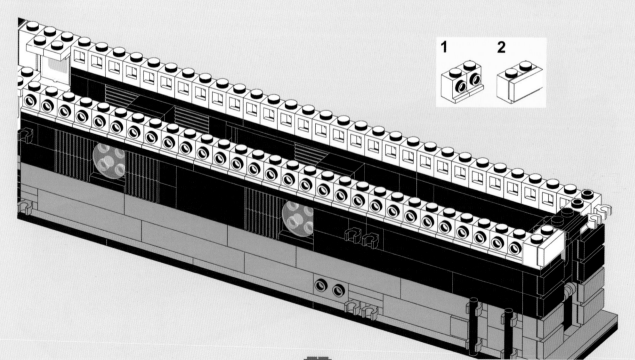

43

1x 1x 1x

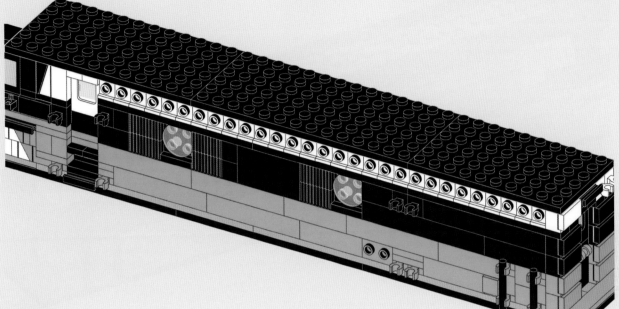

44

2x 1x 1x 1x 2x

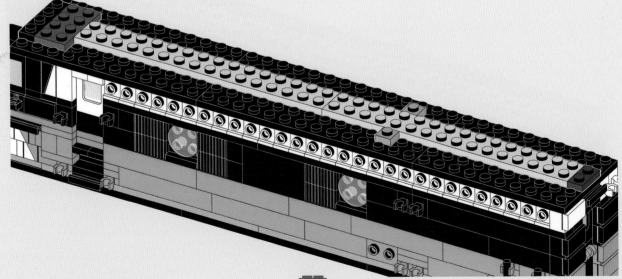

45

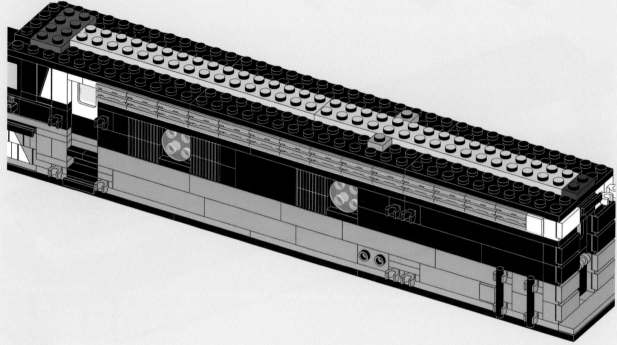

46

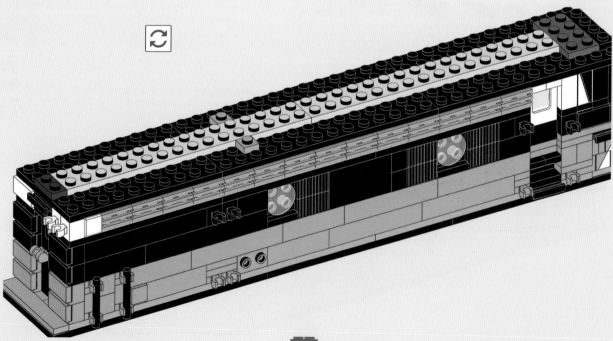

47

1x 1x 2x 10x

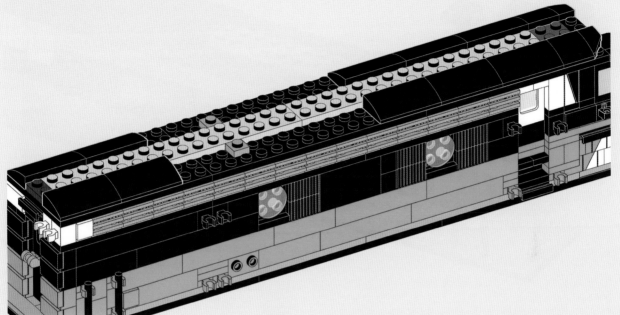

48

2x 3x 2x 1x 2x 1x 2x

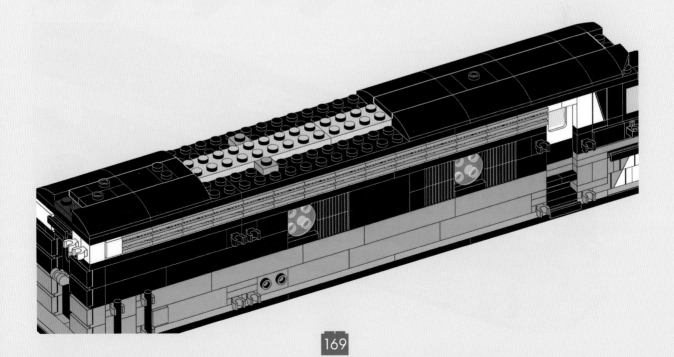

49

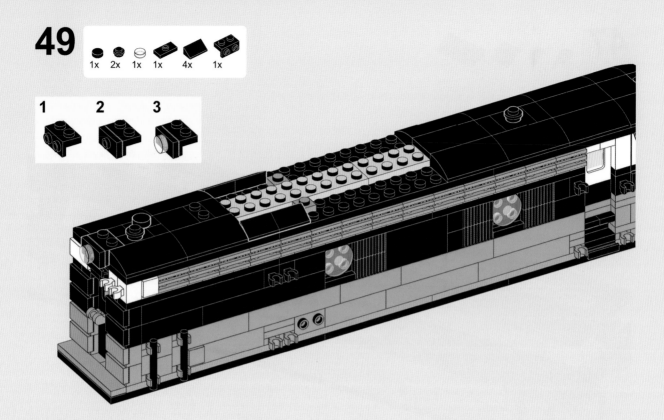

50

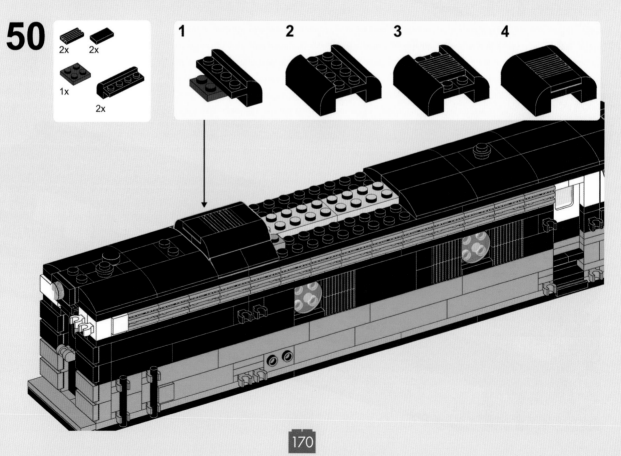

51

1x 2x 4x

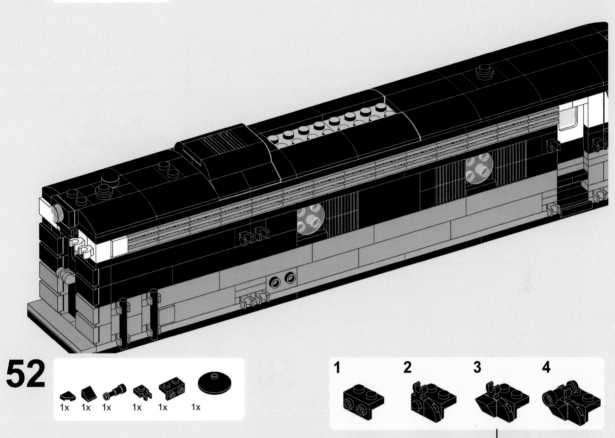

52

1x 1x 1x 1x 1x 1x

1 2 3 4

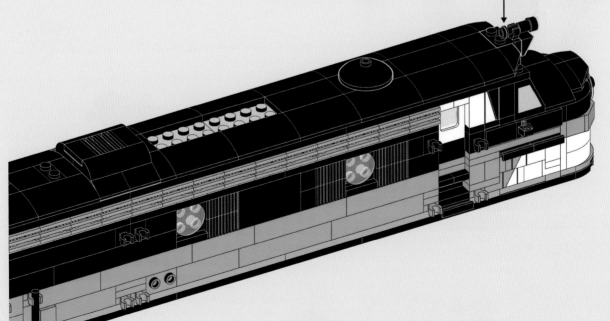

53

3x   1x

54

4x

55

3x

56

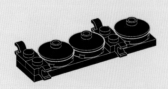

3x

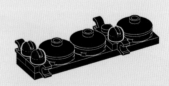

57

1x

58

2x

59

6x

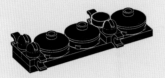

60

2x

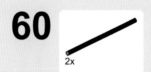

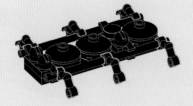

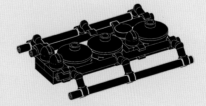

61

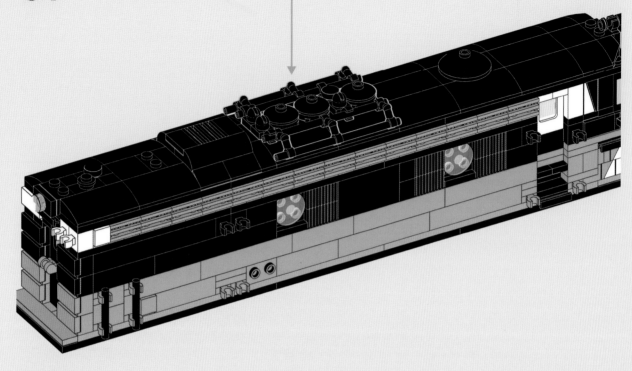

62

1 2 3

2x 2x 2x

63

2x

64
1x 2x

65
2x

66
1x 2x 2x 1x

1 2 3 4

67
4x 2x 2x

1 2 3

2x

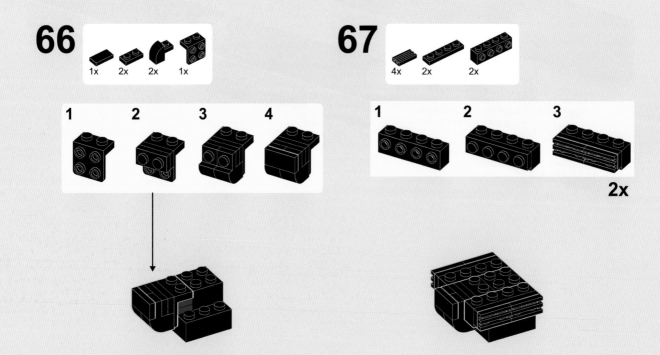

68

69

70

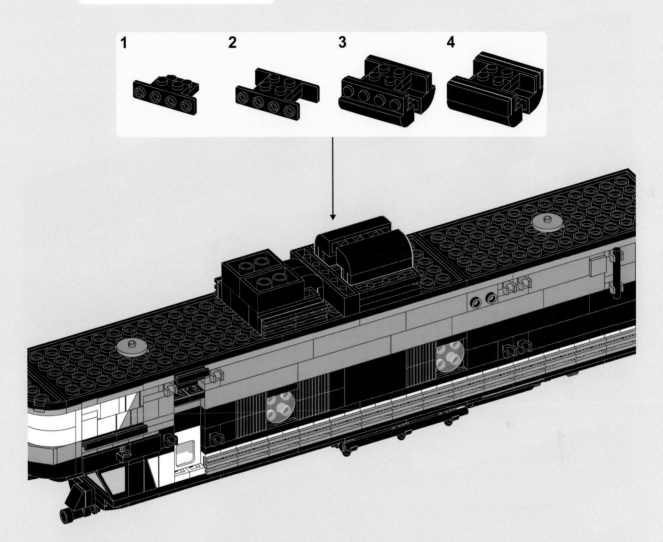

71

72

73

2x   1x

74

4x   4x   2x

1   2   3

2x

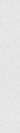

75

1x   1x   2x

76

4x   4x   2x

1   2

2x

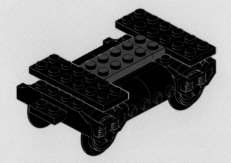

77

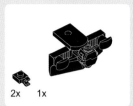

2x 1x

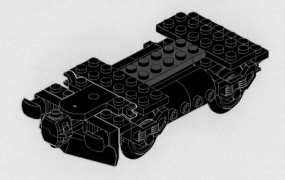

78

4x 1x

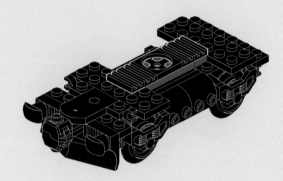

79

2x 2x 3x

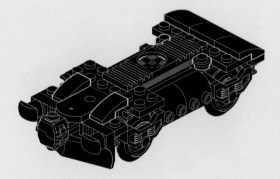

80

2x 2x

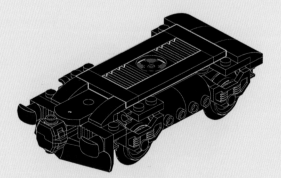

81

1x   1x

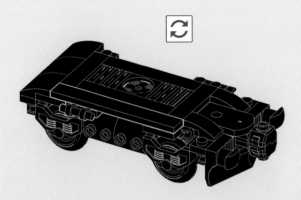

82

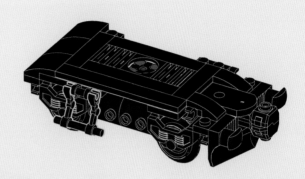

1x   2x

83

1x   1x

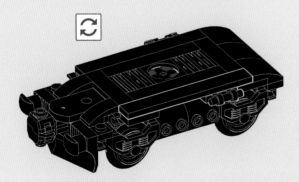

84

1x   2x

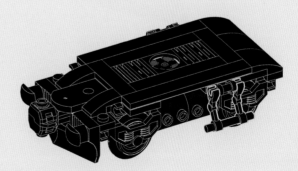

85

86

3x   1x

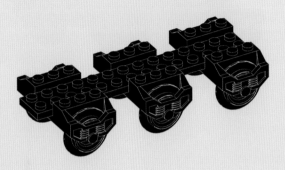

87

4x 2x 2x 4x 2x 4x 2x

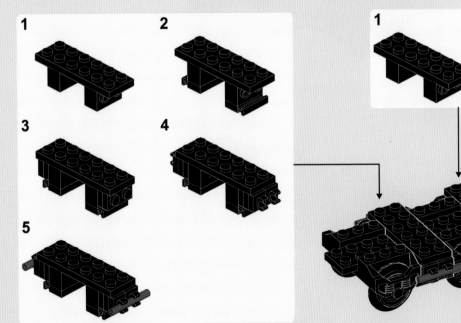

1

2

3

4

5

1

2

88

1x 2x

89

2x 2x 2x 1x

1   **2**   **3**

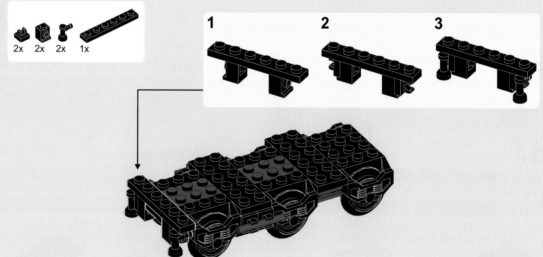

90

2x

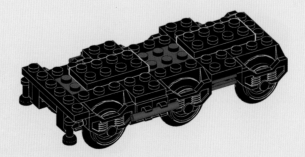

91

2x 2x 1x 1x

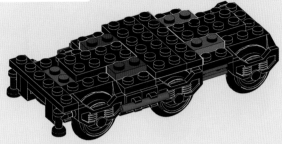

92

3x 2x

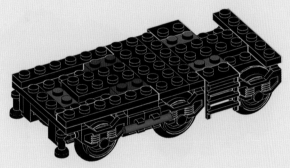

93

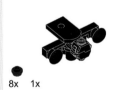

8x 1x

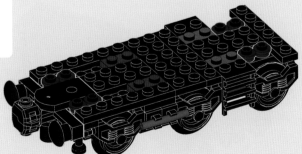

94

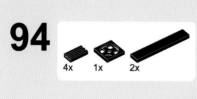

4x   1x   2x

95

2x   4x

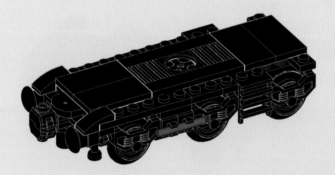

96

4x   2x   2x

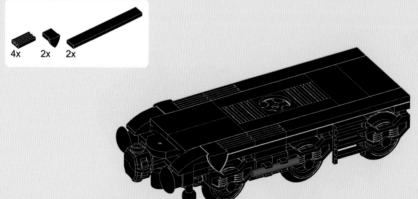

98

1x 4x

99

100

2x

EMD FL9 BILL OF MATERIALS

Item ID	Description	Color	Quantity
30377	Arm Mechanical, Battle Droid	Black	4
93609	Arm Skeleton, Bent with Clips (Horizontal Grip)	Black	6
87994	Bar 3L (Bar Arrow)	Black	6
63965	Bar 6L with Stop Ring	Black	8
99781	Bracket 1×2 - 1×2	Black	2
2436b	Bracket 1×2 - 1×4 with Rounded Corners	Black	4
44728	Bracket 1×2 - 2×2	Black	1
3956	Bracket 2×2 - 2×2 with 2 Holes	Black	4
3005	Brick 1×1	Black	6
3004	Brick 1×2	Black	8
3010	Brick 1×4	Black	2
3009	Brick 1×6	Black	1
3008	Brick 1×8	Black	8
3005	Brick 2×2	Black	5
2357	Brick 2×2 Corner	Black	4
3002	Brick 2×3	Black	2
6005	Brick, Arch 1×3×2 Curved Top	Black	2
30241b	Brick, Modified 1×1 with Clip Vertical (open O clip) - Hollow Stud	Black	8
4070	Brick, Modified 1×1 with Headlight	Black	14
32952	Brick, Modified 1×1×1 2/3 with Studs on 1 Side	Black	2
2877	Brick, Modified 1×2 with Grille (Flutes)	Black	12
11211	Brick, Modified 1×2 with Studs on 1 Side	Black	4
6091	Brick, Modified 1×2×1 1/3 with Curved Top	Black	2
30414	Brick, Modified 1×4 with 4 Studs on 1 Side	Black	2
2653	Brick, Modified 1×4 with Groove	Black	2
30165	Brick, Modified 2×2 Curved Top with 2 Top Studs	Black	4
6081	Brick, Modified 2×4×1 1/3 with Curved Top	Black	4
4740	Dish 2×2 Inverted (Radar)	Black	3
43898	Dish 3×3 Inverted (Radar)	Black	1
44300	Hinge Tile 1×3 Locking with 1 Finger on Top	Black	2
75c07	Hose, Rigid 3mm D. 7L / 5.6cm	Black	2
75c08	Hose, Rigid 3mm D. 8L / 6.4cm	Black	4
4592	Lever Small Base	Black	3
64644	Minifigure, Utensil Telescope	Black	1
4865b	Panel 1×2×1 with Rounded Corners	Black	2

Item ID	Description	Color	Quantity
87552	Panel 1×2×2 with Side Supports - Hollow Studs	Black	1
14718	Panel 1×4×2 with Side Supports - Hollow Studs	Black	2
3024	Plate 1×1	Black	9
4477	Plate 1×10	Black	6
3023	Plate 1×2	Black	11
3710	Plate 1×4	Black	3
3666	Plate 1×6	Black	7
2445	Plate 2×12	Black	1
91988	Plate 2×14	Black	1
3022	Plate 2×2	Black	5
3021	Plate 2×3	Black	1
3020	Plate 2×4	Black	9
3795	Plate 2×6	Black	5
3034	Plate 2×8	Black	1
3033	Plate 6×10	Black	1
3028	Plate 6×12	Black	2
3456	Plate 6×14	Black	3
61252	Plate, Modified 1×1 with Clip Horizontal (thick open O clip)	Black	6
4085d	Plate, Modified 1×1 with Clip Vertical - Type 4 (thick open O clip)	Black	1
15573	Plate, Modified 1×2 with 1 Stud with Groove and Bottom Stud Holder (Jumper)	Black	7
32028	Plate, Modified 1×2 with Door Rail	Black	9
2540	Plate, Modified 1×2 with Handle on Side - Free Ends	Black	2
4175	Plate, Modified 1×2 with Ladder	Black	2
87580	Plate, Modified 2×2 with Groove and 1 Stud in Center (Jumper)	Black	1
4073	Plate, Round 1×1	Black	4
18980	Plate, Round Corner 2×6 Double	Black	1
61409	Slope 18 2×1×2/3 with 4 Slots	Black	2
54200	Slope 30 1×1×2/3	Black	1
85984	Slope 30 1×2×2/3	Black	4
11477	Slope, Curved 2×1	Black	8
15068	Slope, Curved 2×2	Black	5
88930	Slope, Curved 2×4×2/3 with Bottom Tubes	Black	14
24309	Slope, Curved 3×2	Black	2
3660	Slope, Inverted 45 2×2	Black	2
4599b	Tap 1×1 without Hole in Nozzle End	Black	2
3070b	Tile 1×1 with Groove (3070)	Black	4
3069b	Tile 1×2 with Groove	Black	13

Item ID	Description	Color	Quantity
2431	Tile 1×4	Black	4
6636	Tile 1×6	Black	4
4162	Tile 1×8	Black	4
3068b	Tile 2×2 with Groove	Black	2
87079	Tile 2×4	Black	6
15712	Tile, Modified 1×1 with Clip - Rounded Edges	Black	10
35463	Tile, Modified 1×1 with Tooth / Ear Vertical, Triangular	Black	2
2412b	Tile, Modified 1×2 Grille with Bottom Groove / Lip	Black	18
98138	Tile, Round 1×1	Black	2
25269	Tile, Round 1×1 Quarter	Black	1
18674	Tile, Round 2×2 with Open Stud	Black	4
64424c01	Train Buffer Beam with Sealed Magnets - Type 1	Black	1
64415c01	Train Buffer Beam with Sealed Magnets and Plow	Black	1
2878c02	Train Wheel RC Train, Holder with 2 Black Train Wheel RC Train and Chrome Silver Train Wheel RC Train, Metal Axle (2878 / 57878 / x1687)	Black	5
3680	Turntable 2×2 Plate, Base	Black	2
29120	Wedge 2×1 with Stud Notch Left	Black	2
29119	Wedge 2×1 with Stud Notch Right	Black	2
60032	Window 1×2×2 Plane, Single Hole Top and Bottom for Glass	Black	1
6567c01	Windscreen 2×6×2 Train with Trans-Clear Glass	Black	1
3023	Plate 1×2	Blue	5
3710	Plate 1×4	Blue	1
3022	Plate 2×2	Blue	3
3004	Brick 1×2	Dark Bluish Gray	2
3023	Plate 1×2	Dark Bluish Gray	3
3021	Plate 2×3	Dark Bluish Gray	2
3020	Plate 2×4	Dark Bluish Gray	2
85861	Plate, Round 1×1 with Open Stud	Dark Red	8
30145	Brick 2×2×3	Light Bluish Gray	1
22885	Brick, Modified 1×2×1 2/3 with Studs on 1 Side	Light Bluish Gray	2
87552	Panel 1×2×2 with Side Supports - Hollow Studs	Light Bluish Gray	4
87544	Panel 1×2×3 with Side Supports - Hollow Studs	Light Bluish Gray	1
3024	Plate 1×1	Light Bluish Gray	2
4477	Plate 1×10	Light Bluish Gray	6
3023	Plate 1×2	Light Bluish Gray	2
3795	Plate 2×6	Light Bluish Gray	2
26047	Plate, Modified 1×1 Rounded with Handle	Light Bluish Gray	1

Item ID	Description	Color	Quantity
15573	Plate, Modified 1×2 with 1 Stud with Groove and Bottom Stud Holder (Jumper)	Light Bluish Gray	2
3070b	Tile 1×1 with Groove (3070)	Light Bluish Gray	2
3069b	Tile 1×2 with Groove	Light Bluish Gray	1
63864	Tile 1×3	Light Bluish Gray	2
6636	Tile 1×6	Light Bluish Gray	1
2412b	Tile, Modified 1×2 Grille with Bottom Groove / Lip	Light Bluish Gray	28
27925	Tile, Round Corner 2×2 Macaroni	Light Bluish Gray	2
3679	Turntable 2×2 Plate, Top	Light Bluish Gray	2
99780	Bracket 1×2 - 1×2 Inverted	Orange	2
3005	Brick 1×1	Orange	12
3010	Brick 1×4	Orange	8
3009	Brick 1×6	Orange	10
4070	Brick, Modified 1×1 with Headlight	Orange	20
87087	Brick, Modified 1×1 with Stud on 1 Side	Orange	2
11211	Brick, Modified 1×2 with Studs on 1 Side	Orange	2
3024	Plate 1×1	Orange	12
3023	Plate 1×2	Orange	11
3666	Plate 1×6	Orange	3
2420	Plate 2×2 Corner	Orange	2
4085d	Plate, Modified 1×1 with Clip Vertical - Type 4 (thick open O clip)	Orange	16
54200	Slope 30 1×1×2/3	Orange	4
28192	Slope 45 2×1 with Cutout without Stud	Orange	2
3070b	Tile 1×1 with Groove (3070)	Orange	2
3069b	Tile 1×2 with Groove	Orange	4
3068b	Tile 2×2 with Groove	Orange	2
75c03	Hose, Rigid 3mm D. 3L / 2.4cm	Red	4
3021	Plate 2×3	Red	2
3020	Plate 2×4	Red	2
41677	Technic, Liftarm 1×2 Thin	Red	2
15712	Tile, Modified 1×1 with Clip - Rounded Edges	Red	2
2412b	Tile, Modified 1×2 Grille with Bottom Groove / Lip	Red	2
4282	Plate 2×16	Tan	2
99780	Bracket 1×2 - 1×2 Inverted	White	2
3005	Brick 1×1	White	4
6005	Brick, Arch 1×3×2 Curved Top	White	4
30241b	Brick, Modified 1×1 with Clip Vertical (open O clip) - Hollow Stud	White	2
4070	Brick, Modified 1×1 with Headlight	White	60

Item ID	Description	Color	Quantity
87087	Brick, Modified 1×1 with Stud on 1 Side	White	2
32952	Brick, Modified 1×1×1 2/3 with Studs on 1 Side	White	2
3024	Plate 1×1	White	4
3023	Plate 1×2	White	6
61409	Slope 18 2×1×2/3 with 4 Slots	White	2
54200	Slope 30 1×1×2/3	White	4
3070b	Tile 1×1 with Groove (3070)	White	2
3069b	Tile 1×2 with Groove	White	1
3068b	Tile 2×2 with Groove	White	1
60032	Window 1×2×2 Plane, Single Hole Top and Bottom for Glass	White	2
3023	Plate 1×2	Yellow	2
3710	Plate 1×4	Yellow	1
4740	Dish 2×2 Inverted (Radar)	Trans-Clear	1
60601	Glass for Window 1×2×2 Flat Front	Trans-Clear	3
2654	Plate, Round 2×2 with Rounded Bottom (Boat Stud)	Trans-Clear	4
98138	Tile, Round 1×1	Trans-Clear	1

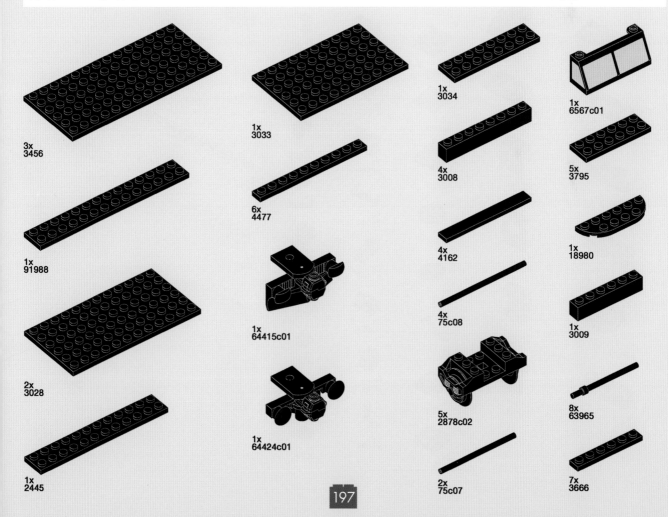

3x
3456

1x
91988

2x
3028

1x
2445

1x
3033

6x
4477

1x
64415c01

1x
64424c01

1x
3034

4x
3008

4x
4162

4x
75c08

5x
2878c02

2x
75c07

1x
6567c01

5x
3795

1x
18980

1x
3009

8x
63965

7x
3666

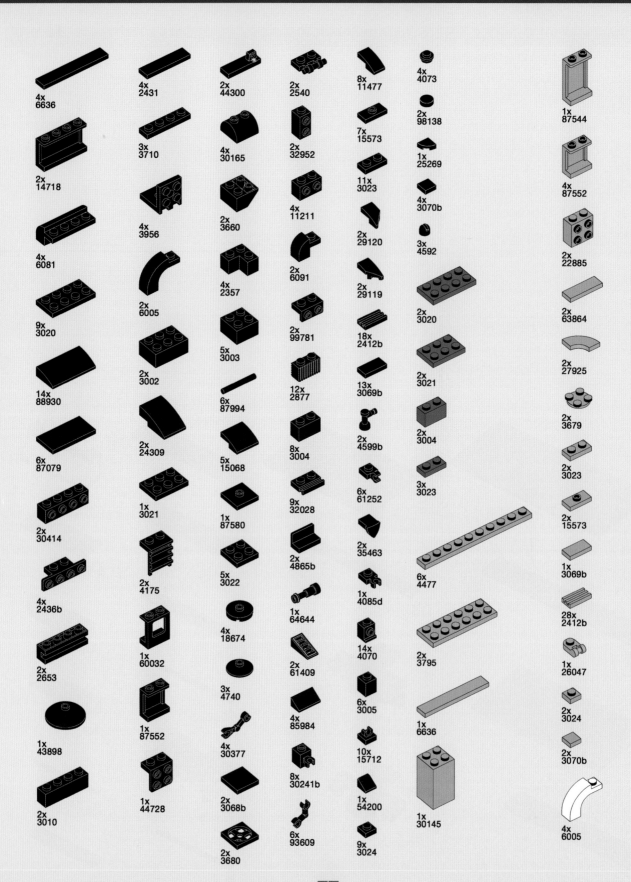

4x 6636
4x 2431
2x 44300
2x 2540
8x 11477
4x 4073
1x 87544

2x 14718
3x 3710
4x 30165
2x 32952
7x 15573
2x 98138
4x 87552

4x 6081
4x 3956
2x 3660
2x 32952
11x 3023
1x 25269
4x 22885

9x 3020
2x 6005
4x 2357
4x 11211
2x 29120
4x 3070b
2x 63864

14x 88930
5x 3003
2x 6091
2x 29119
3x 4592
2x 27925

6x 87079
2x 3002
6x 87994
2x 99781
18x 2412b
2x 3020
2x 3679

2x 30414
2x 24309
5x 15068
12x 2877
13x 3069b
2x 3021
2x 3023

4x 2436b
1x 3021
1x 87580
8x 3004
2x 4599b
2x 3004
2x 15573

2x 2653
2x 4175
5x 3022
9x 32028
6x 61252
3x 3023
1x 3069b

1x 43898
1x 60032
4x 18674
2x 4865b
2x 35463
6x 4477
28x 2412b

2x 3010
1x 87552
3x 4740
1x 64644
1x 4085d
2x 3795
1x 26047

1x 44728
4x 30377
4x 85984
14x 4070
1x 6636
2x 3024

2x 3068b
8x 30241b
6x 3005
1x 30145
2x 3070b

2x 3680
6x 93609
10x 15712
4x 6005

1x 54200
9x 3024

198

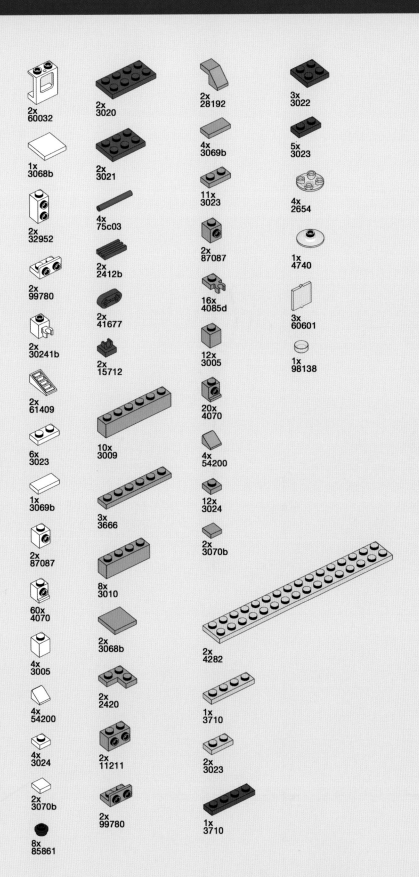

2x
60032

1x
3068b

2x
32952

2x
99780

2x
30241b

2x
61409

6x
3023

1x
3069b

2x
87087

60x
4070

4x
3005

4x
54200

4x
3024

2x
3070b

8x
85861

2x
3020

2x
3021

4x
75c03

2x
2412b

2x
41677

2x
15712

10x
3009

3x
3666

8x
3010

2x
3068b

2x
2420

2x
11211

2x
99780

2x
28192

4x
3069b

11x
3023

2x
87087

16x
4085d

12x
3005

20x
4070

4x
54200

12x
3024

2x
3070b

2x
4282

1x
3710

2x
3023

1x
3710

3x
3022

5x
3023

4x
2654

1x
4740

3x
60601

1x
98138